Concrete and Plastic

Environmental Cultures Series

Series Editors:
Greg Garrard, University of British Columbia, Canada
Richard Kerridge, Bath Spa University, UK

Editorial Board:
Frances Bellarsi, Université Libre de Bruxelles, Belgium
Mandy Bloomfield, Plymouth University, UK
Lily Chen, Shanghai Normal University, China
Christa Grewe-Volpp, University of Mannheim, Germany
Stephanie LeMenager, University of Oregon, USA
Timothy Morton, Rice University, USA
Pablo Mukherjee, University of Warwick, UK

Bloomsbury's *Environmental Cultures* series makes available to students and scholars at all levels the latest cutting-edge research on the diverse ways in which culture has responded to the age of environmental crisis. Publishing ambitious and innovative literary ecocriticism that crosses disciplines, national boundaries, and media, books in the series explore and test the challenges of ecocriticism to conventional forms of cultural study.

Titles available:
Anthropocene Realism, John Thieme
Bodies of Water, Astrida Neimanis
Cities and Wetlands, Rod Giblett
Civil Rights and the Environment in African-American Literature, 1895–1941, John Claborn
Climate Change Scepticism, Greg Garrard, George Handley, Axel Goodbody, Stephanie Posthumus
Climate Crisis and the 21st-Century British Novel, Astrid Bracke
Cognitive Ecopoetics, Sharon Lattig
Colonialism, Culture, Whales, Graham Huggan
Contemporary Fiction and Climate Uncertainty, Marco Caracciolo

Digital Vision and Ecological Aesthetic, Lisa FitzGerald
Doing Animal Studies with Androids, Aliens, and Ghosts, David P. Rando
Ecocollapse Fiction and Cultures of Human Extinction, Sarah E. McFarland
Ecocriticism and Italy, Serenella Iovino
Ecospectrality, Laura A. White
Environmental Cultures in Soviet East Europe, Anna Barcz
Fuel, Heidi C. M. Scott
Imagining the Plains of Latin America, Axel Pérez Trujillo Diniz
Literature as Cultural Ecology, Hubert Zapf
Reading Underwater Wreckage, Killian Quigley
The Living World, Samantha Walton
Nerd Ecology, Anthony Lioi
The New Nature Writing, Jos Smith
The New Poetics of Climate Change, Matthew Griffiths
Radical Animism, Jemma Deer
Reclaiming Romanticism, Kate Rigby
Teaching Environmental Writing, Isabel Galleymore
This Contentious Storm, Jennifer Mae Hamilton
The Tree Climbing Cure, Andy Brown
Weathering Shakespeare, Evelyn O'Malley

Forthcoming titles:
Ecocriticism and Turkey, Meliz Ergin

Concrete and Plastic

Thinking through Materiality

Kylie Crane

BLOOMSBURY ACADEMIC
LONDON • NEW YORK • OXFORD • NEW DELHI • SYDNEY

BLOOMSBURY ACADEMIC
Bloomsbury Publishing Plc, 50 Bedford Square, London, WC1B 3DP, UK
Bloomsbury Publishing Inc, 1359 Broadway, New York, NY 10018, USA
Bloomsbury Publishing Ireland, 29 Earlsfort Terrace, Dublin 2, D02 AY28, Ireland

BLOOMSBURY, BLOOMSBURY ACADEMIC and the Diana logo are trademarks of
Bloomsbury Publishing Plc

First published in Great Britain 2024
This paperback edition published 2026

Copyright © Kylie Crane, 2024

Kylie Crane has asserted her right under the Copyright, Designs and Patents Act,
1988, to be identified as Author of this work.

For legal purposes the Acknowledgements on pp. ix–x constitute an extension of this
copyright page.

Cover design: Rebecca Heselton
Cover image © "We Were Here Now" (2018) mosaic ©
by Julie Sperling. (Photograph courtesy of Conan Stark)

This work is published open access subject to a Creative Commons Attribution-
NonCommercial-NoDerivatives 4.0 International licence (CC BY-NC-ND 4.0, https://creative
comm ons.org/licen ses/by-nc-nd/4.0/). You may re-use, distribute, and reproduce this work
in any medium for non-commercial purposes, provided you give attribution to the copyright
holder and the publisher and provide a link to the Creative Commons licence.

Bloomsbury Publishing Plc does not have any control over, or responsibility for, any
third-party websites referred to or in this book. All internet addresses given in this
book were correct at the time of going to press. The author and publisher regret any
inconvenience caused if addresses have changed or sites have ceased to exist,
but can accept no responsibility for any such changes.

A catalogue record for this book is available from the British Library.

Library of Congress Cataloging-in-Publication Data

Names: Crane, Kylie, author.
Title: Concrete and plastic : thinking through materiality / by Kylie Crane.
Description: First edition. | New York : Bloomsbury Publishing PLC, [2024] |
Series: Environmental cultures ; vol. 31 | Based on the author's
thesis, written in German, under the title Habilitation. |
Includes bibliographical references and index.
Identifiers: LCCN 2023042528 (print) | LCCN 2023042529 (ebook) |
ISBN 9781350380592 (hardback) | ISBN 9781350380639 (paperback) |
ISBN 9781350380608 (pdf) | ISBN 9781350380615 (ebook)
Subjects: LCSH: Concrete–Philosophy. | Plastics–Philosophy. |
Concrete–Environmental aspects. | Plastics–Environmental aspects. |
Plasticity–Philosophy. | Object (Philosophy)
Classification: LCC TA439 .C69 2024 (print) | LCC TA439 (ebook) |
DDC 620.1/92301–dc23/eng/20240224
LC record available at https://lccn.loc.gov/2023042528
LC ebook record available at https://lccn.loc.gov/2023042529

ISBN: HB: 978-1-3503-8059-2
PB: 978-1-3503-8063-9
ePDF: 978-1-3503-8060-8
eBook: 978-1-3503-8061-5

Series: Environmental Cultures

Typeset by Newgen KnowledgeWorks Pvt. Ltd., Chennai, India

For product safety related questions contact productsafety@bloomsbury.com.

To find out more about our authors and books visit www.bloomsbury.com
and sign up for our newsletters.

Contents

List of Figures		viii
Acknowledgements		ix
1	Introduction	1
2	Plastic Pacific	53
3	Megadam materialities	83
4	Unpacking plastic	115
5	Concrete ruins	147
6	Conclusion	179
Bibliography		191
Index		209

Figures

1.1	Berlin Wall souvenir	2
2.1	'Plastiglomerate' by Joshua Franzos	54
3.1	'Boulder Dam, 1942' by Ansel Adams	87
3.2	'Boulder Dam, 1941' by Ansel Adams	89
3.3	'Matilija Dam, California 2014' by Carsten Meier	114
4.1	'Surrounded Islands, Biscayne Bay, Greater Miami, Florida, 1980–83' by Christo and Jeanne-Claude	127
4.2	'The Floating Piers, Lake Iseo, Italy, 2014–16' by Christo and Jeanne-Claude	130
5.1	'三芝飛碟屋 – panoramio (1)' by The Erica Chang	165

Acknowledgements

I acknowledge the multiple privileges that meant I could write this book. I acknowledge the systematic disadvantages that are un-dis-entangle-able from my own privileges. I acknowledge the violently displaced and decimated peoples, the Noongar, whose lands sustained my childhood. I acknowledge the livelihoods and livelinesses that sustain my life now. I acknowledge the shortcomings of these acknowledgements.

I acknowledge that a project like this ekes out space and time (ten years!) from other things, and other people. I owe thanks, at least, to many things and people.

This book had a previous life as a thesis, a German creature called a *Habilitation*, submitted at the University of Potsdam. For their generous reports I would like to thank Anja Schwarz, Nicole Waller and Ingrid Hotz-Davies. I would like to thank the colleagues and friends who stood by me during that phase and at a distance, outside, to celebrate.

Thanks go to my colleagues at the University of Mainz (Germersheim), at the University of Potsdam and at the University of Rostock: in particular to Oliver Czulo and Sabina Matter-Seibel in Germersheim, Anja Müller-Wood in Mainz, for her remarks on chapters, as well as her support for the project in its earlier stages, and at Potsdam, I would like to thank Lars Eckstein (in addition to those mentioned above), the people of the RTG 'minor cosmopolitanisms', and the very collegial English Department there, as well as the fine people of my current workplace in Rostock.

Thanks go to further colleagues who were generous in giving their time to read and comment on parts of the project. My thanks go to Roman Bartosch and Sonja Frenzel, and to Ingrid Hotz-Davies, who continues to be a most perceptive mentor. Nina Jürgens deserves a special thanks for her readings and comments and for the index. Thanks must go to many other colleagues and students who replenished my thinking with seemingly random concrete and plastic stuff, and to Baldeep Kaur for being a great reading partner. For invites to give talks about different parts of this project, thanks go to Roman Bartosch, Joanna Durczak, Sonja Frenzel, Alexandra Ganser, Kerstin Knopf, Pavan Malreddy, Kai Merten, Timo Müller, Miriam Nandi, Caroline Rosenthal, Wolfgang Struck and the organizers and participants at several GAPS summer schools. My thanks also go to Alexandra Rossiyskaya for her work on helping shape the final manuscript.

At Bloomsbury and adjacent, I would like to thank Greg Garrard (series editor), Ben Doyle (Publisher), Laura Cope (Assistant Editor), Joe Skingsley (publishing team), Rebecca Heselton (cover design) and Balasuwathiga (Project Manager at Newgen). For their feedback, I would like to thank the anonymous reviewers from Bloomsbury. I would also like to explicitly thank Julie Sperling for allowing me to use a photo of her mosaic 'We Were Here Now' on the cover.

I am obliged to the University of Rostock and its University Library, whom I would like to thank for their financial support making Open Access publication possible.

Thanks go to the artists who have allowed me to include their artwork in this book, and thanks also to the thinkers, writers and makers whose work made my work possible: You will find these people acknowledged throughout the book.

I want to also thank my friends, who have provided support over the years: the time span has meant several moves, several new jobs and two children. Here I think in particular of Mathilde Bessert-Nettelbeck, Julia Boll, Julie Fielder, Jennifer Henke, Nina Jürgens, Ines Koenen, Louise Niggermeier, Angy Polania, Geoff Rodoreda, Anja Schwarz, Julia Spanke, and the ex-bowling-crew in Bremen.

Thanks go to my family. To my mum, Ann Crane, RIP, to my dad, Les Crane, and to my brothers, Murray Crane and Gavin Crane. Thanks also to Reinhild Duda and Thomas Müller for their support. And, of course, to Sebastian, Elliot and Oscar Duda, for their patience and other things. This is the book I've been writing for so long.

1

Introduction

Denkstoff: A souvenir

An object on my desk: A strip of plastic, 2 mm thick, curved like a 'C', and fixed in it, a chunk of concrete, around 30 mm × 20 mm × 15 mm in size. The concrete part has yellow and orange paint on one, flattened, side, otherwise its edges are uneven. The plastic has printed on it a stylized map of Berlin, with flags showing the division into sectors following the Second World War, and reads 'Berlin Wall; 13.09.1961–09.11.1989; Allied Checkpoint Charlie' on the side where the paint on the concrete is visible.[1] The other side reads 'You are leaving the American Sector' in English, Russian, French and, in much smaller print, 'Sie verlassen den amerikanischen Sektor'.

Describing it now in such detail several years after purchasing it on a whim – feeling 'tickled' that such a small object in a souvenir stand might make use of both of the materials in my book – it becomes clear to me that the object itself does not explicitly claim to be what I never believed it was: a piece of the Berlin Wall. The sum of the materials – in particular the concrete, the paint and, crucially, the quasi-preserving plastic – together with the information printed on the plastic obliquely suggest that this is what I am purchasing. But the artefact stops short of the outright claim to such a provenance. My souvenir does not re-present the Berlin Wall, but it cannot exist without it.

Despite, or perhaps because of, its banality, it is an interesting object. And it is an interesting amalgamation of materials. As an object, it relies heavily on its context: as a souvenir bought at the Berlin airport, through the writing and images printed on the plastic, and through the lurid yellow and orange paint on one, flatter, surface of the concrete. As souvenir, it functions within a

---

[1] As of 5 February 2018, the wall was down longer than it was up.

 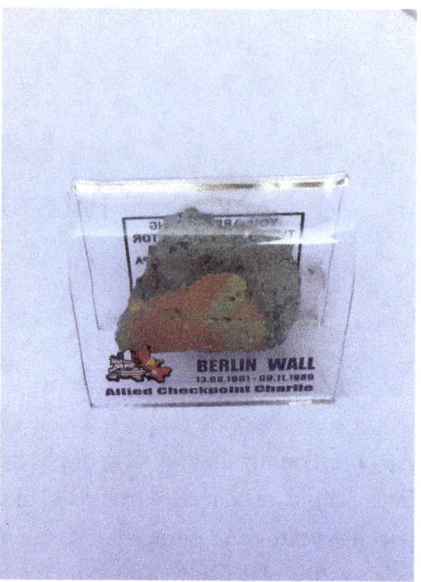

Figure 1.1 Berlin Wall souvenir.
L: With sunlight and shadow, atop concrete. R: Less 'busy' background. Both photographs by the author.

semiotics of travel and tourism (cf. Urry 2002, Culler 1988: 153–67). According to the dictionary, it is 'a thing that is kept as a reminder of a person, place or event' ('Souvenir'). It is a material reminder of a trip to Berlin, for myself, and perhaps it also reminds others that I am the sort of person who has been to Berlin. The souvenir also speaks to political projects, specifically the historical circumstances that gave rise to the construction of the wall. In this, it also speaks to the ways in which we organize and conceptualize the (social) world within which we live.[2] In particular ways, the object, sitting on my desk, *objects* to being 'simply' an object.

---

[2] The trajectory from the singular 'I' to the plural 'we' is never straight or straightforward. I use the 'we' in this book to evoke a collective position that extends beyond the singular of the 'I' throughout this book. I recognize that this is problematic for the ways in which it interpellates the reader into positions of identification *or* resistance. The 'we', nonetheless, is used for exactly this reason, and with my own hesitations riding with it. This 'we' should work to probe my own privilege as well as, potentially, that of my readers: This privilege – although it will map differently over the identity categories of gender, sex, sexuality, class, nationality, residence, disability, ethnicity and race – is a given. Those who can and who take the time to read this book inhabit positions of privilege: privileges of education, of time (affordances of occupation or preoccupation), of class and of intellectual capacity, amongst others. I also use these first-person pronouns to indicate the positions and angles (cf. esp. Ahmed 2010: 37), which the material of the book presents. You(!) will note there are a series of footnotes trying to do this work throughout the book: My hope is that, together, they can trace the problems, and also the necessity, of the first-person plural.

For Bill Brown, the 'object-ness' of objects is a function of the way in which they are *made* objects, that is, placed in relation to subjects (humans, often of an unspecified kind), where 'things' are that which trouble, or escape, such simple, dichotomic relations. He argues: 'As they circulate through our lives, we look *through* objects (to see what they disclose about history, society, nature, or culture – above all, what they disclose about *us*), but we only catch a glimpse of things' (Brown 2001: 4, emphasis in original). Things, or 'thingness', begin(s) to become evident when such relations break down: 'The story of objects asserting themselves as things', Brown claims, 'is the story of a changed relation to the human subject and thus the story of how the thing really names less an object than a particular subject-object relation' (ibid.).

The 'clunkiness' of the souvenir as an object is its 'thingness', the ways it forms and resists 'straightforward' attributions of meaning.[3] Such 'straightforwardness' is, of course, a function of very specific ways of thinking. For Alison Jones and Te Kawehau Hoskins, in their discussion of Māori ontology, a 'thing' 'is not understood as discrete or independent, but emerges through, and as, relations with everything else' (Jones and Hoskins 2016: 80). This is one example for how indigenous epistemologies recognize relations in ways with which dominant epistemologies might struggle, and part of that has to do with language. In my examination of the souvenir, the object (noun) begins to object (verb) at the moment of discovery: I didn't buy it believing it was a piece of the Berlin Wall – even as I later realized that the object itself does not specifically claim to be so. But it doesn't 'stop' here: 'souvenir', deriving from the Latin *subvenire*, 'occur to the mind', is incredibly generative, lots of 'stuff to think about'. A kind of *Denkobjekt*, a *Denkfigur*, but most pressingly, *Denkstoff*. It has, for me, never been a 'simple object'.

The appeal of the souvenir was not only the way that it pushed back on its meanings, the way that it purported to be something it probably wasn't. Simultaneously, the myth of clear relations between it as object and its social relations (that somehow are external to it) crumbled, like concrete. For me, the

---

[3] Things assert themselves most readily when they no longer function as objects, for Bill Brown. He explains his 'thing' as encompassing 'a massive generality as well as particularities, … the *concrete* yet ambiguous within the everyday …[,] a place holder for some future specifying operation …[,] an amorphous characteristic or a frankly irresolvable enigma' (Brown 2001: 4, emphasis added). Things, Brown argues, tend to 'index a certain limit or liminality, to hover over the threshold between the nameable and unnameable, the figurable and the unfigurable, the identifiable and unidentifiable' (ibid.: 5). Note the auspicious use of 'concrete' in the first quotation, and the 'plasticity' of the ways in which 'things' shift through thresholds.

appeal of the object was also the materials from which it was made, the substance as much as its substantiality: its materiality.

The materiality of the souvenir is its very concreteness and plasticity: made of concrete (noun) and plastic (noun), but also in the other senses of the words. It 'exist[s] in a material or physical form' (as adjective; 'Concrete') and is 'easily shaped or moulded' (again, as adjective; 'Plastic'). This duality of meanings stretches the relations of the materials to their respective environments, and such stretching reaches through dimensions of place and time, that is, it is temporal and spatial. In this book, I shift from asking questions of the object-thing-item-artefact – that which pertains to be discrete, somehow – to the materials of which it is made.

Take a second look at the photos of my souvenir (Figure 1.1). I took them outside, to take advantage of the good light conditions. I used my phone to take the left photograph first, only to realize how difficult it was to see it against the background of my balcony. The patterns of the background 'interrupted' the image. I went inside and got a piece of paper, altered the orientation of the souvenir to be parallel to the camera and shifted it into the shade. This was a 'better' image, I initially thought. But: It is only better with respect to some criteria.

Tim Ingold notes the exclusion of particular phenomena of our worlds from philosophical considerations. He invites his reader in 'Materials against Materiality' to consider sunlight as part of the 'material world', despite its exclusion from conceptual frames (cf. Ingold 2007: 3–4). In the images of my souvenir, the left photograph is the more captivating, perhaps not just because of the 'interferences' of the concrete tiles upon which it was photographed, but also the inclusion of the shadows, giving 'shape' and mass to the object. The shadow appeals to the senses: not just the visual sense, but also the tactile ones (as flitting between direct sunlight and the shade will prove, on a hot sunny day as much as a cold one). It seems as if reckoning with artefacts as objects, if not things, relies on a visual paradigm of knowledge, that 'draws boundaries' around objects: The shadow impinges on this otherwise seemingly absolute separability.

Other senses will also undoubtedly suggest boundedness: We can touch, smell and taste. We run into things, drop things, feel things; we can listen to them, sniff at them (or turn away from their odours), and savour their flavours, or spit things out in disgust. But this list already suggests the ways in which this boundedness, or separability, becomes (increasingly) fragile. Sight and sound require a medium for us to perceive them, and this medium links us to that which is perceived; touching, smelling and tasting are even more 'trans-corporeal' senses, that is, those, following Stacy Alaimo, 'in which the human is always inter-meshed with the more-than-human world' (Alaimo 2010: 2).[4] Smell, like taste, involves particles passing into our bodies.

My wonder, happenchance-like-happiness (cf. Ahmed 2010), at the souvenir is coupled with a dismay that derives from realizing the purchase has solidified the demand for what is ultimately a piece of junk, a-thing-becoming-waste. These oscillating feelings unsettle the presence of the souvenir on my desk. Stephanie Bunn helps here, reminding me to think about the act of making as a 'working *with* rather than a doing *to*' (Bunn 1999: 15, emphasis in original): How can we work *with* the souvenir, the *Denkstoff*, rather than doing something *to* it? My answer is to consider it in terms of its materiality, the materials of which it is made. It is a turn towards material culture that takes the *material* seriously.

'Materiality' I use in a way that could also be 'material-ness': it references the matters of material. It is not used in any sense that is divorceable from materials themselves. In the frame of this book, Tim Ingold's statement that urges that 'we take a step back, from the materiality of objects to the properties of materials' (Ingold 2007: 9) does not make sense.[5] My thinking insists on relations between sites, peoples, times, biotic and abiotic matters, and the ways these are upheld or truncated. If Ingold feels secure in stating that 'it is significant that studies of so-called material culture have focused overwhelmingly on processes of consumption rather than production' (ibid.), this critique is met in the following sections, where I trace the becomings of plastic to 'get' concrete.

Why concrete *and* plastic? The materials that this book works through are both modern, at least in their application: they derive from industrial processes. They are abundant, ubiquitous, and readily identifiable (in opposition to, say, silica): they are (also) 'cheap' (cf. Patel and Moore 2020). I would wager that most (probably all) of my readers will be able to identify several items comprising these materials in their near vicinities, or at least in their day-to-day lives. Concrete and plastic *accompany* us through our lives.

---

[4] Alaimo's explication of the 'trans-corporeal' continues by stressing movement and shifts: 'By emphasizing the movement across bodies, trans-corporeality reveals the interchanges and interconnections between various bodily natures. But by underscoring that *trans* indicates movement across different sites, trans-corporeality also opens up a mobile space that acknowledges the often unpredictable and unwanted actions of human bodies, nonhuman creatures, ecological systems, chemical agents, and other actors' (Alaimo 2010: 2, emphasis in original).
[5] Whilst there are some aspects of Ingold's chapter that make coherent points for this book – 'Wherever life is going on, they [materials] are relentlessly on the move – flowing, scraping, mixing and mutating' (Ingold 2007: 11) for instance – the contradistinctions and oppositions Ingold establishes to make his point elsewhere read awkwardly in thinking through materialities here, possibly due to his dependencies on phenomenology, which often problematically draws on assumptions of universalist, privileged perception.

And yet, crucially, they are not dormant nor static. In production, they transform through various states (being fluid, then formed, then forms). Both plastic and concrete can be described as 'paradigmatic matter-like material[s]' that 'when poured into a rigid formwork ... appear [to] seamlessly to take on the shape of [their] mould' (Lloyd Thomas 2015: 272).[6]

Concrete and plastic are present on different scales. Concrete, for instance, in household decorations (fashionable at the time of drafting this introduction) and even jewellery, as well as the material in memorials, (large) buildings and dams, infrastructures; plastic as 'microscopic' or tiny particles, both in our bodies and outside (as in glitter or body peelings), as well as packaging material, and as insulation of pipelines, also infrastructures. The kinds of connections I am forging through an examination of the materials belie such scales, going beyond the readily apprehensible 'near-human' scales of the objects, to the molecular (through hormonal interventions) and the global (through trade). As Waters et al. suggest, polychlorinated biphenyls are 'among the many distinct geo-chemical signatures that human activities have introduced into the sedimentary record' (Waters et al. 2016: web). In *Pollution is Colonialism*, Max Liboiron is more insistent, noting bisphenol A, or BPA's 'ubiquity is better understood as violence, a manifestation of the permission-to-pollute system that allows BPA to be found in nearly all those tested' (Liboiron 2021: 87). The extremes of this scale – the molecular and the global – are in fact more interlinked than immediately apparent at the intermediate levels of perception (see also Chapter 2).

The fluidity of concrete and plastic at certain stages of their production lends itself to thinking about the interconnections of matter, at both discernible and indiscernible scales. That, further, these materials interact with fluids – the water contained by dams (cf. Chapter 3) or the oceans within which they coalesce (cf. Chapter 2) – is more than a coincidence: The fluidity of materials needs to be conceptualized *alongside* their quotidian status as objects that appear to resist, condense or suspend fluidity.

Whilst we might consider them as *being already formed* (as a consequence of having been poured, moulded or printed) and somehow inert, dormant or passive, concrete and plastic rather *continually shape* our lives and our imaginations. This, too, happens at various scales: global, local, individual and cellular. Plastic and concrete shape our (imagination of our) future and our

---

[6] For Katie Lloyd Thomas, they are part of a longer tradition of discussions of matter in the hylomorphic mode, comprising bronze, wax, clay, as well as gold (cf. Lloyd Thomas 2015: 271, 282–4). 'These materials work well as paradigms of matter because, at least in their liquid state, they are formless in themselves', she argues (ibid.: 271).

responses to ever-shifting presences and the ever-shifting present. Although the bulk of this book separates the materials into analytical chapters – Chapters 2, 3, 4 and 5 – the impetus of reading the two materials in conjunction produces numerous insights, resulting in concrete plastics and plastic concretes. This is the *plasticoncrete* of the next section.

The emphasis on concrete and plastic arises from a conviction that the materials of the world are not 'simply' shaped by us[7] – and the capacity for both materials to *be formed* addresses this aspect – but to query the ways in which, conversely, materials of the world might shape us. It is a project, then, that finds a particular urgency in its articulation within the rubrics of the Anthropocene, which I discuss in more detail later, in the section 'Condensing relations'. Thinking the Anthropocene demands careful consideration of the impact of human inventions and interventions on the environment and the way that this shapes us and our temporally and spatially removed others. The Anthropocene maps unevenly through disciplines, time, materials and space. It is not a concern that can be compartmentalized to the future, but might be, for a privileged few, a concern that is not considered immediate.

Against, or with, this, consider my conflicting feelings about the enduring materiality of the souvenir. As what I call a 'future artefact', my piece of concrete wrapped in a 'c' of plastic will be, in a speculative future, only material. The concrete and plastic souvenir of (or at least, 'pertaining to') the Berlin Wall is going to be thrown 'away' at some time in the future: It has a future dimension which emerges *with* its materiality. It invites the invention of stories – what will happen to it? Who will throw it 'away'? – not only of the future, but also of the past: Was it ever part of the Wall? Where was it made? Where do its 'bits' come from? And, what of my 'Berlin Wall' souvenir when memory of the wall no longer is in circulation? What of the physicality of this piece of concrete fused with plastic when the clues to its provenance have disappeared (rubbed off, or the languages have disappeared)? I respond to such questions by thinking about 'future artefacts', evoking disruptive potentials that unfold through transformations of multiple presents and presences.

Without the cultural context of the Berlin Wall, and without the broader cultural context that gave rise to my purchase, my souvenir doesn't fulfil any particular function. It's actually pretty useless: It can't contain or preserve anything (Chapter 4), it does not generate power (Chapter 3). If it does not

---

[7] Again, recognize this first-person plural as a shorthand for a more-than-me plural, speculative as it is, and imbued with positions of privilege.

'signify' within a cultural context that values souvenirs, and that understands the importance of the Berlin Wall, then it is rather useless. The German slang word is *Stehrumski* (a thing that stands [*steh*] around [*rum*]), in English perhaps it is a thingimibob: a thing, exceeding its function as an object, by retreating back from its functionality. But inevitably, irrevocably, indisputably: material.

My souvenir, the *Denkstoff*, the piece of concrete wrapped in plastic, is undoubtedly a fabricated object. It was produced somewhere. Plastic is an industrial product, as is cement, a key component of concrete. Thinking materiality pushes back on the timelines of production, however, extending beyond the shaping into the prehistories of materials. Thinking materials connects plastic bottles to prehistoric creatures (e.g. Morton 2010: 29).[8] It also connects coral reefs with concrete high-rises, as evoked in the strange, unsettling juxtapositions in Romesh Gunesekera's novel *Reef*: 'Mister Salgado only slowed down when we came to the skull-heaps of petrified coral – five-foot pyramids beside smoky kilns – marking the allotments of a line of impoverished lime-makers, tomorrow's cement fodder, crumbling on the loveliest stretch of the coast' (Gunesekera 1994: 59–60). Here, off the coast of an imagined Sri Lanka, we recognize the material relations of concrete, with impoverished workers chipping away at localized beauty for construction sites elsewhere.

The temporal dimension of the futurity of my *Denkstoff* is an affair of the imagination. It is an invention (see also 'Invention: Coming into relations' section). Spatially removed dimensions are, for most of us, accessed through the imagination, as in the Gunesekera vignette. It is in the dimensions that stretch from the specific manifestations of the materials, connecting them to disparate places and times, that two, equally crucial, objectives and specifications of this project emerge: The cultural use of materials, coupled with recognition of the ways in which materials are formed by us and also form us (and thus shape our ways of engaging with the world).

Cultural representations, as human-made artefacts, might seem to willingly reproduce the anthropocentric conceptualization of materials and materiality (if distinct) that this book pushes back on: They are, predominantly, representations made by humans and for humans. However, this does not mean that they might not develop, or *invent*, alternate ways of imagining the world and its relations. Such cultural representations, or 'texts', do not have to be overtly didactic (in

---

[8] Although not the dinosaurs Morton suggests, see Greg Garrard's critique: 'While it sounds cool to say that "We drive around using crushed dinosaur parts" (p. 29), the fact is that oil comes mainly from marine micro-organisms' (Garrard 2012: 201). I still like the idea of thinking about this with dinosaurs, though.

the 'prescribing' or even 'patronizing' sense) to engage and challenge their audiences' perceptions of the world: Oftentimes it is rather the subtle ways in which they shift or alter the way in which their audiences reflect on their own positionings and behaviours.[9]

Not just a framework, culture is seen as product *and* producing agent, as creation, creator and creating (cf. also Sheller 2014: 64).[10] How, from a perspective schooled in literary and cultural studies, is it possible to interpret ('read') the objects, things, indeed materials of the environment as not hyperseparated (Plumwood 1993) from representation, but as crucially, irrevocably, tangibly entangled with representation? Rejecting the commitment to mimesis that pervades assumptions that the ontological world is radically different and yet simultaneously grounds for (politically motivated schools of) literary analysis, I propose patterns of interpretation that diffract rather than reflect (drawing on the work of Karen Barad). Accordingly, texts are not readily divided into primary and secondary, a division that appears to suggest that some texts merely 'support' interpretations of others, or provide a structural framework through which they are interpreted. Following Jussi Parikka, we might consider that the (Foucauldian) archive 'is not only about the statements and rules found in books and libraries. Instead [perhaps: Moreover], it is to be found in technological networks of machines and institutions, patterns of education and drilling' (Parikka 2015: 2). To supplement this list, this sense of the archive is also in the unwritten records of oral histories (to name an obvious omission), as well as in the practiced and rehearsed (learnt and taught) practices of interaction with the material world passed on through trades, crafts and 'tinkering'.

---

[9] I adopt the term 'positioning' to reckon with shifting allegiances in manners pertaining to identity politics. Stuart Hall, in 'Cultural Identity and Diaspora', puts forward this term in opposition to essence, as a means of negotiating practices that nevertheless make meaning possible (1990). It is strategic, also in the sense of Gayatri Spivak's 'strategic essentialism' (2003), but also entails arbitrary and contingent endings producing an excess, something 'left over'. The term 'positioning' references the spatial domain, as do the close-cognates 'situated knowledge' (Haraway 1988), 'angle' (Ahmed 2010: 37) and 'besides' (Sedgwick 2003: 8). The gerund form serves to stress this as a process, an ongoing endeavour, with the further, desired, effect of flitting between stasis and motion. It thus picks up on Karen Barad's critique of self-positioning as a reflective practice that ultimately remains in the domain of mirroring, or, as literary scholars would put it, mimesis (e.g. Barad 2007: 72). Why do I stress positioning? As a corrective to universalist tendencies masked as 'subject' positions, and to stress the ethical impetus of this book.

[10] A basic tenet of this book is that the forms available for the description of the world influence the way that access to this world is mediated and understood. This is not to suggest a teleology of increased comprehension, a development of ideas that reach towards a singular and attainable truth, but, instead, I want to suggest that asking different questions through different means with different texts will pluralize the truths of the world, and jostle beside other, perhaps more canonical or established, epistemologies.

Thinking concrete, thinking plastic: *Plasticoncrete*

For this section I use a neologism,[11] *plasticoncrete*, to probe the ways in which the concrete informs the plastic, the ways in which concrete pushes back on plastic, and vice versa.[12] *Plasticoncrete* is a heuristic device, used to bring the two terms as close together as possible. Thinking concrete, thinking plastic – thinking *plasticoncrete* – means dwelling on relations forged between the two materialities: their etymologies and forms, their temporal and spatial locations, their import in quotidian material practices. *Plasticoncrete* entails mobilizing meanings with materialities.

The key terms of this book – concrete and plastic – are used both as a noun and as an adjective in English. Historically, the use of concrete and plastic as adjectives preceded the respective uses as nouns: In both cases, this suggests that the material, the noun, cannot be divorced from the associations of the adjective. Thinking concrete, thinking plastic entails mobilizing the valencies of *plasticoncrete* in language: the conceptual infrastructure that gives rise to (material) practices.

Concrete and the concrete: Fiona Allen notes that 'as soon as concrete acquires a definite article, "the concrete", it becomes an abstraction' (Allen 2015: 237). This is concrete becoming less concrete, becoming abstract. At the same time, Leonard Koren observes: 'We rely on [concrete's] absolute physical integrity to lend solidity and certitude, real and metaphoric, to our lives' (Koren 2012: 13). This is concrete becoming (more? is that possible?) concrete.

In etymological terms, concrete has Latin roots, deriving from *concrescere*, to 'grow together'. Its historical use was as a descriptive word, as a 'quality belonging to a substance', before being used to refer to 'nouns embodying attributes' (Chantrell 2002: 111). It is in this sense that we might say that something is 'set in concrete' meaning that it is fixed and unalterable. In another related sense, concrete means something specific or definite, as in to have 'concrete plans'. It is also used as an adjective to suggest a material object in opposition to an abstract quality, state or action. In this way, it is possible to

---

[11] In this book, I refrain from neologisms on the whole, preferring to tease out relations and ideas in the forms they present themselves – and the forms they work to resist.
[12] The term *plasticoncrete*, further, forges a relation to 'plastiglomerate', a substance formed with plastic, but with stones as an aggregate: a concrete form of plastic. See the 'Condensing relations' section later for more on this substance, in particular regarding its role as a marker for the Anthropocene.

say 'concrete obstacles' for objects that are, in fact, not made out of concrete, but that are not 'abstract'. Concrete slips between its various meanings.¹³

The shifting references become evident in the quote from Leonard Koren's introduction to *Concrete*, where the adjectival use maps onto the associations with the material. And as Rowan Bailey notes in 'Concrete Thinking for Sculpture', '*concrete* not only describes *how* it becomes form, it also expresses the character of its fixed condition' (Bailey 2015: 242, emphasis in original). This is the double valency crucial to this book: concrete as materiality, and as quality.¹⁴

Plastic similarly slips between noun and adjective, although this movement produces different effects. Deriving from the Greek *plassein* 'to mould', via Latin and French, the historical use of plastic, which emerged in the mid-seventeenth century, was to denote a quality, specifically: 'characteristic of moulding' (Chantrell 2002: 381). This is evident in phrases that denote a scope of creativity (e.g. 'the medium is rather plastic'), pertaining to moulding or modelling in three dimensions (e.g. 'the plastic arts'), or in biology (referencing a capacity for adaptation).¹⁵ The noun 'plastic' draws on the adjectival associations to denote a material that is shape-able and formable. More recent use of the adjective plastic draws on associations with the material. This is the sense of the 'plastic smile': not a smile made *of* plastic, but one considered inauthentic.

Jeffrey L. Meikle, accordingly, argues that a 'glance at almost any work of fiction of the late 1960s or 1970s [in the United States], whether popular or literary, reveals that plastic – as material and as concept – had become a pejorative used with confidence that the reading public would not object' (Meikle 1995: 262). Further, he suggests that 'everyone understood a reference to a "plastic person" or a "plastic smile"' (ibid.: 263). Putting aside his confidence that his interpretation will be the same as any reader's ('everyone'), associations with plastic do seem to have become increasingly negative, a trajectory that has continued beyond the

---

[13] In translation, these relations shift. Many other Indo-European languages use the term *beton/Beton*. It is related to 'bitumen' (asphalt) and derives from the Old French for mud, gravel and rubbish (cf. Forty 2012: 11, and also Braun and Pfeifer 1989: 160). Concrete, in these languages, becomes entangled with sediments and rubbish, forging a further etymological relation with concerns of the Anthropocene (see the section 'Condensing relations: Materialities in the Anthropocene').

[14] Concrete denotes a composite material. It is comprised of cement, aggregate (usually stones and sand) and water. As 'reinforced concrete', concrete also entails steel, which provides greater structural stability.

[15] German uses *Plastik* to denote sculptures (as well as the material). As Jürgen Eichhoff observes, the distinction between *Plastik* and *Plaste* has been used to make this distinction more obvious (Eichhoff 1980: 163, fn 35). Although the latter, *Plaste*, is less broadly used today, it was quite common in the German Democratic Republic (sometimes called East Germany), where Eichhoff suggests it was a shortened form of *Plastmasse*. The prefix *Plast-* derived from the English term (ibid.: 165–6). German also uses the term *Kunststoff* (*Kunst*: art, artificial, *Stoff*: stuff, material).

time frame he designates. A sense of this meaning can be traced through pop songs which oftentimes equate 'plastic' with 'fake'.

Thinking with plastic will, for most of us now, be negative. But plastic is more complicated than that. The abundancy of plastic across the globe today comes from somewhere, and can, I think, be coupled to early celebratory stances towards the material. The figure of 'Plastic Man' is symptomatic for this, which Meikle describes as the 'apotheosis of conservative plastic utopianism' (ibid.: 68). The figure emerges in an essay by Victor E. Yarsley and Edward G. Couzens which appeared in *Science Digest* in 1941 and was also published as part of a volume entitled *Plastics* (also 1941). The authors elicit a world populated by plastic objects, aiding development and mitigating risk. They write:

> Let us try to imagine a dweller in the 'Plastic Age' that is already upon us. This creature of our imagination, this 'Plastic Man', will come into a world of colour and bright shining surfaces, where childish hands find nothing to break, no sharp edges or corners to cut or graze, no crevices to harbour dirt or germs … surrounded on every side by this tough, safe, clean material. (Yarsley and Couzens 1941: 154)

In this imagined world, plastic is pervasive – the material of crockery, hygiene, clothes and stationery, and also of the surfaces of the environment at home and school: 'The manufacturer of the future will say, not 'of what material shall I make this article?' but "what kind of plastic shall I use?"', Yarsley and Couzens suggest (ibid.: 157). The prophetic tone resonates: Yarsely and Couzens's predictions are eerily accurate, and they conclude their volume by noting:

> When the dust and smoke of the present conflict have blown away and rebuilding has well begun, science will return with new powers and resources to its proper creative task … the perfect expression of the spirit of planned scientific control, the Plastics Age. (ibid. 158)

For Meikle (writing some fifty years after Yarsley and Couzens's publication) and for the contemporary reader (now more than seventy years later), the 'utopia' of 'Plastic Man' is not entirely unrecognizable.[16] Plastic's malleability – its capacity to take on numerous forms and functions – *materializes* in its ubiquity and engenders both marvel and dismay.

* * *

---

[16] Adrian Forty notes that accounts of concrete usually begin with the Romans 'and their discovery of naturally occurring pozzolanic cements' (Forty 2012: 7); instead, he starts with the observation that Thomas More's *Utopia* has a description of concrete, specifically of the houses of the Utopians.

Plastic is a chameleon materiality.¹⁷ Petrochemical-based plastics are, and I will return to this point throughout the book, fossils made manifest in manufactured forms. Before the first synthetic plastics were produced, substances occurring in nature – gutta-percha, shellac and the horns of animals – were used as plastic material. Bakelite was invented in 1907, the first plastic based on a synthetic polymer. It was moulded into thousands of forms, such as cases for radios, telephones and clocks, and billiard balls. After the Second World War, improvements in chemical technology, political shifts towards decolonization (and hence reduced access to the natural substances) as well as plastic manufacturers looking to develop markets for the materials previously purchased as part of military expenditure led to an explosion in new forms of plastics – among them polypropylene and polyethylene. These rapidly found commercial application in a wide spectrum of products, from coffee cups, to shampoo bottles, to bags, eyeglass frames, medical instruments and, well, almost everything, including all kinds of things you didn't even know you could need.

Roland Barthes thus observes in the chapter 'Plastic' from *Mythologies*: 'More than a substance, plastic is the very idea of its infinite transformation; as its everyday name indicates, it is ubiquity made visible' (Barthes 2000: 97). Plastic is, to Barthes, 'the first magical substance which consents to be prosaic' (ibid.: 98) and, further, 'it is precisely ... this prosaic character [that provides] ... a triumphant reason for its existence: for the first time, artifice aims at something common, not rare' (Barthes 2000: 98). The ideas of malleability and ubiquity shape-shifting into ideas of magic is not restricted to Barthes's (well-known) account, nor are they only applicable to plastic.

In *My Cocaine Museum*, Michael Taussig traces the Roman story of cement, suggestive for the ways in which it 'makes stones seem alive and capable of amazing metamorphoses' albeit only 'once they have been processed by man

---

¹⁷ For the purposes of this book, plastic is primarily considered a derivative of fossil sources, predominantly petrochemicals. Indeed, for Charles Moore and Cassandra Phillips, 'It's axiomatic. Where there is oil there is plastic' (Moore and Phillips 2012: 23). 'Organic' plastics, or bioplastics, manufactured from plants rather than coal and oil derivatives, are quicker to biodegrade and are definitely quicker to 'grow'. But, bioplastics are not necessarily 'more environmentally friendly' than petrochemical plastics: The intensive agricultural practices entailed in growing the raw materials for bioplastics (e.g. the use of fertilizers, extensive monocultures), the displacement of other agricultural projects (i.e. growing plastic instead of food), the capacity of bioplastics to interrupt recycling of petrochemical plastics and, as a last example, the release of methane when bioplastic products end up in landfill all complicate any straightforward replacement of petrochemical plastics (cf. Cho 2017). The forms of their affectual chemistry differ from those of petrochemical plastics, which is why they are sometimes not 'thought with' the overall arch of the book. As also becomes evident elsewhere in this book, replacement narratives are hardly simple substitutions, but rather intricate entanglements. Bioplastics are not so different from fossil fuel–derived synthetics in this respect.

[*sic*]' (Taussig 2004: 162). First-century Roman architect and builder Vitruvius, Michael Taussig remarks, 'wanted a substance like stone but malleable. When you stop to think about it, this is like something out of a fairy tale: *like stone but malleable*' (ibid.: 161, emphasis in original, cf. also Miodownik 2014: 74). For Taussig, the magic of concrete is an intrinsic quality *of* its technology: 'You start with stone. You make a powder. And then in the process of building, you add water and end up with a new form of 'stone' in accord with the shape desired. It sounds like magic but we call it technology' (Taussig 2004: 161). Taussig's shorthand of 'magic' is suggestive of a process with a surprising end-result, which can, however, be emulated. 'Magic' also is suggestive of the ways in which the processes of concrete and plastic remain elusive to us in our everyday practices. Recognizing this means recognizing concrete and plastic as cultural practices, that probe, prohibit and produce wonder.[18]

Certainly, the 'magic' of concrete and plastic is a wonder of technology, itself both scientific and wonderous: Jussi Parikka, for example, suggests that 'even the mundane is produced through a mix of the archaic underworld and the refined scientific process' (Parikka 2015: viii). The particular processes of concrete (particularly: cement) and plastic manufacture remain strange and foreign and are 'outsourced' (see also 'Invention: Coming into relations'). In his book *Concrete and Culture*, Adrian Forty asserts that 'concrete can be more accurately described as a *process* than as a *material*' (Forty 2012: 44, emphasis in original). In a consideration of concrete as process, the 'growing together' of the etymology becomes evident. For, concrete continues *becoming* concrete after the pour. This is sometimes rendered visible and tangible in the patterns of feet, paws and tyres, or scratched initials, in concrete slabs. The processes of curing concrete are also susceptible to atmospheric conditions. It cures for roughly four weeks at which stage it reaches 'full workable strength' (everreadymix.co.uk), but forms new bonds continuously. There is, thus, nothing particularly *concrete* about concrete. Only within the constraints of particular definitions – such as the above-mentioned 'full workable strength' – can any notion of the concreteness of concrete be entertained.

Concrete is a process situated in place: As Forty notes, 'One cannot take a piece of reinforced concrete, a "sample", to show what the structure will be made of, for reinforced concrete only happens when the work is cast and the network of forces between steel and concrete becomes "live"' (Forty 2012: 51). Weather

---

[18] I have developed these ideas in more detail in an analysis of Karen Tei Yamashita's *Through the Arc of the Rain Forest* (cf. Crane 2016).

conditions are also important. The interplay of time and place, the processes of 'becoming concrete', are crucial to all kinds of concrete (reinforced or not). The processual characteristic of concrete is one that can be asserted for plastic, too.[19] As Jennifer Gabrys suggests: 'Plastics are materials in process; they fragment and break down, while also generating new material arrangements' (Gabrys 2013: 208). As processes, these materials are relations set in action.

Plasticoncrete as process means considering that (these) materials might appear to be stable ('concrete') but that they are also always shifting ('plastic'). This is a function of scale, both temporal and spatial. The dimension of space patterns through the materialities at scales, both discernible and not: The scale of the object, through which most of our encounters with concrete and plastic occur, is only one of these. Thinking materialities evokes the scale of the molecular as well as the scales of globalization and logistics, and, with the temporal, the scales of sourcing, outsourcing and effect/affect. The dimension of time, further, diffracts these latter scales, evoking the long duration of oil, for instance, as a component of plastic, or the sources of lime for cement.

* * *

Both concrete and plastic are ubiquitous, and are present in masses – sometimes *as* masses – and are produced at industrial scales.[20] Their ubiquity is, amongst other factors, a consequence of their 'cheapness', where 'cheapness' is understood as a mechanism that masks hidden but nonetheless accumulating 'costs'. These 'costs' include those of planned obsolescence, resulting in continued production and the accelerating accumulation of materials and goods; toxic relations and Wasteocenes through which the '(re)production of wasted people and places serves the purpose to create a safe and worthy "we"' (Amiero 2021: 31), that is, the externalization of the materials (and the concomitant risks of production and decay) to other places and other times; and other messy or truncated relations of extractive economies (cf. also Patel and Moore 2020).

Concrete and plastic are both fluid before they are solid. They flow into forms before they flow into the realms of the day-to-day. The insight that *plasticoncrete* is fluid-come-solid also generates the converse insight: solids becoming less solid (if not fluid). For plastic, this means recognizing that chains of polymers – the molecular scale of plastics – emerge in more recent accounts as insidious insertions into the endocrine and hormonal levels of

---

[19] Some plastics are reinforced, too.
[20] With concrete, it is the cement that is produced industrially. Concrete itself is created *in situ*, an aspect that is important to other points I make here.

our beings. Concrete thinking recognizes that concrete is not as stable as it purports, and particularly not as immobile as it might seem. Concrete begins as a liquid, before it is set, and must be kept in motion in order to prevent it from hardening before it has been poured into its mould. It is often doubly mobile on its way from production site to construction site, in the drums of slowly rotating concrete mixers on the backs of trucks moving through cities and other concrete spaces.

The cement vital to concrete is comprised of crushed and fired limestone. Cement is inherently stable materials (rocks) made instable (calcium silicates) to be made stable again (concrete), and something of Taussig's transformative magic adheres to my understanding of these processes. The resultant powder is mixed in with water to create a gel, a semi-solid, more familiar to most of us in the form of toothpaste or jellies and jams. In cement, the processes that follow the adding of water create a skeleton comprised of calcium silicate hydrate fibrils, which Miodownik explains are 'chrystal-like entities that grow from the calcium and silicate molecules' (Miodownik 2014: 68). This process continues, and the fibrils will bond not only with each other but also with any sand or small stones (the aggregate) added to the mixture, and, in reinforced concrete, to steel. Rachel Harkness, Cristián Simonetti and Judith Winter harness gerunds in their subtitle ('Gathering, Flattening, Curing'), thus, to be indicative of the processual nature of working *with* concrete. Concrete, then, not as 'concrete', but as process, as a growing together.

In the ancient applications of concrete, what 'comes together' was also 'blown apart', as volcano ash was used for cement in Ancient Roman times. Fossilized, or calcified, stone is a crucial component of plastic, *and* the limestone that gives rise to concrete's cement. And it is one that patterns through 'stone simulacra' (concrete) and 'stone oil' (eventually: plastic): Stone harbours the long scales of geological change, fossil-records of past organic life.

Concrete and plastic scarcely shift from fluid into solid forms to remain as such for 'eternity': They are not dormant, docile or passive. Instead, their materiality continues to assert itself, along the long scales of their presence. Thinking plastic brings Topshop 'Mom Jeans' with plastic knee inserts into conversation with band-aids and prostheses, CDs speak to Coke bottles, eyeglasses get tangled with medical face masks. It is not so much a dialogue-like dialectics, probably more a heated debate: more than one voice competing for attention, and interruptions, analepses, changes in tone, speed and pitch abound. Thinking concrete, similarly, might bring Tadao Ando's 'Meditation Space' artwork into conversation with the cover to a glossy art book, for example, *Concrete* (Hall

2012) – a paper book about concrete covered in paper with an image of a concrete wall, itself formed by timber buttresses – via the shell on the door at Le Corbusier's 'Notre-Dame-du-Haut in Ronchamp', and the highway overpass outside. And even some *plasticoncrete* examples: plastic houses (and 'printed' concrete houses), concrete lamp fittings and asphalt made of recycled plastic. The *material* world is full of interesting, innovative and inimical ways of coming into relations with materiality.

Invention: Coming into relations

In *Stuff Matters*, Mark Miodownik's stories narrate the 'discovery' of modern materials. Both in his 'Fundamental' chapter, where he explores concrete, and in his 'Imaginative' chapter, which addresses plastic, these stories take fortunate turns (Miodownik 2014: 64–84, 125–59). 'The Romans', Miodownik starts off this 'concrete' story, 'got lucky with concrete':

> Instead of having to experiment with heating up different combinations of ground-up rock to white-hot temperatures, they found ready-made cement in a place called Pozzuoli just outside Naples. … All the Romans had to do was put up with the smell and mine the rock powder that had been accumulating for millions of years. This naturally made cement is slightly different from modern ('Portland') cement and requires the addition of lime to make it set. But once they had worked this out and added stones for strength, they had in their hands for the first time in human history the fundamentally unique building material that is concrete. (ibid.: 70–1)

In this casual, upbeat story, the use of concrete is a lucky coincidence, and the Romans more or less fell into relations with cement.

Such narratives of happenchance abound in tales of concrete (and plastic, as I will show later). Another example is given by Robert Blezard, in 'The History of Calcareous Cements', which similarly couches the tale of Portland cement as accidental: William Aspdin, the story goes, left the family business of cement manufacturing to join a different plant on the south banks of the Thames in London, where he discovered that 'clinkered or "overburnt" material substantially increased the strength of his cement. As William had only a limited knowledge of chemistry', Blezard argues 'the discovery must almost certainly have been accidental' (Blezard 2004: 8). As Blezard also notes, the 'story of the invention of Portland cement has not been easy to *disentangle*. As has happened so often in chemical history, the real innovator of an industry was not the initial discoverer of a reaction' (Blezard 2004: 8, emphasis added).[21]

In the chapter on plastic, Miodownik is more 'imaginative' about his discovery story, writing it as a 'rudimentary screenplay', albeit interspersed with explanatory notes. In 'Notes for Scene 2', Miodownik points out that 'in the late nineteenth century, the beginning of the golden age of chemical engineering, a growing understanding of chemistry coincided with entrepreneurial opportunities for making money out of the invention of new materials' (Miodownik 2014: 137–8). The example he gives – of celluloid and billiard balls – is paradigmatic for these kinds of stories for a number of reasons: An organic material (ivory) used for a specific purpose (billiard balls) needs replacement (as the source becomes scarcer and demand rises), such that the financial incentive to create – to invent – new materials that fulfil this purpose rise. As 'Scene 4' reveals, this context gives rise to parallel timelines of 'invention', in this case criss-crossing the Atlantic between the United States and the UK, as Alexander Parkes and Daniel Spill make materials they call Parkesine and Xylonite, respectively (cf. ibid.: 141–5).

At the same time, many stories of 'coming into plastic' follow a pattern of the 'accident', of 'chance' or of 'luck', some even stressing the happenchance greatly. Such stories place their emphasis often on the individuals as agents – evoking notions of genius, of people driven by insight – at the same time as eliciting stories of chance. In doing so, such stories often neglect to accord the materials any agency, and they also background contextual matters.

In one such 'discovering plastic' story, Stephen Fenichell mobilizes a number of these tropes – surprise, accident and, curiously, the absence of humans at crucial stages of the process – with respect to viscose:

> Since he [Charles Topham] happened to be building a new lamp factory at Kew, after a day of frustrating work with viscose spinning, he hopped the late train to London to make a *surprise* inspection of the plant site. Returning to Erith two days later, Topham was *shocked* to find the viscose solution he had left behind *dramatically transformed*. Where before it had dropped straight from its jet into a caustic-soda bath – in which it was *supposed* to coagulate like an egg white in boiling water – in a chain of distinct beads, the aged viscose now stretched nearly to its breaking point before finally splitting into fragments under the

---

[21] The stories about modern concrete tend to gather around Portland cement, glossing over other stories. A brief overview of pre-Portland-cement accounts of the 'discoveries' of 'Post-Roman use of cements' can be found in Hendrik G. van Oss's 'Background Facts and Issues Concerning Cement and Cement Data' (2005: 2–3). I return to the notion of 'entanglement' in the section 'Modern materials' later.

accumulated tension of its own weight. Before it split, the monofilament yarn, when drawn out to its maximum length, was quite strong. By yet another *inspired accident*, Topham had *stumbled upon* the value of aging, or 'ripening', viscose. (Fenichell 1996: 123, emphases added)

The phrases I have emphasized stress the accidental tone of the story. Note in particular the phrase 'stumbled upon': Here are materials getting in the way.

Nevertheless, I want to trouble this kind of accidental discovery story. 'Discovery', with its insistence on 'uncovering', insists on the agency of the discoverer, some kind of conduit of revelation. Jens Soentgen's German-language *Stoffgeschichte* – which translates as 'material story' as well as 'material history' – redresses such typical accounts of the invention and discovery with his example of rubber, emphasizing the practices of indigenous peoples of South America. As Soentgen notes, indigenous use of materials is often encoded in terms of the traditional and ritual, rather than being seen as a technological or cognitive achievement (cf. Soentgen 2013: 298). The encoding of knowledge as traditional erroneously suggests a stasis – a sense of inactivity and inalterability – as well as a sense of innateness. Indigenous knowledge is 'shifted to the margins', even 'de-historicised', as much as appropriated in this story. Either the practices that give rise to such material knowledge are ignored or they are appropriated *with* the raw materials, that is, practices of engagement with the materials are subsumed within the status of 'raw material'. With rubber, to return to Soentgen's example, the story traces Charles Macintosh and Thomas Hancock competing for the 'discovery' of vulcanization, the result of which is a rubber that neither melts in the heat nor turns brittle in the cold. Crucially, this is a process that emulates results *already materialized* in the bounciness of the rubber balls brought to Europe from the Americas by Christopher Columbus. The relations which constituted the emergence of practices of vulcanizing rubber were transported *with* the material itself, even as they were 'backgrounded' in the teleology of Western discovery.

Elizabeth Shove et al. note that narratives of discovery emphasize originality and ingenuity for a number of reasons, such as the securing of funding and patents. In these kinds of stories, 'The retrospective retelling of technological trajectories frequently reveals the critical role of prospective narratives within and as an integral part of the innovation process' (Shove et al. 2007: 97). Nevertheless, the backgrounding or disregard of other knowledges or practices is attendant to such narratives. *Discovery* collects stories of exertion of power, of access to political and financial infrastructure, and of the willing backgrounding of knowledges, at the same time as it evokes a linear trajectory of modernity.

In this book, I think of innovative, paradigmatic or otherwise enduring material shifts in terms of *invention*. Invention entails a 'coming into' rather than discovery's 'uncovering'. Invention entails originality in method and goal, but it also stresses the presence of the pre-existing, that is, the practices and materials of the context into which is entered. 'Coming into' thus also entails an 'entering', thus discursively allowing for the habitations and agencies of others. Whilst discovery also suggests the pre-existing (that which is 'hiding' under the cover), it does not stress the active component of relationality, but rather a discrete container which is un-covered. I prefer the term 'invention' to stress the mess of entanglements and dependencies, even contingencies. Invention, in this understanding, shifts agency away from a trajectory of human genius as locus of 'dis-covery' and towards an intra-action of human and material (cf. Barad 2007, and below).[22]

Where some stories of invention often entail entanglements in colonization, in which indigenous practices are 'backgrounded', some such 'replacement narratives' are embroiled with processes of decolonization. DuPont's development of Nylon can be read as an example of such a replacement narrative, where materials are developed to replace scarce or controlled materials. Nylon, initially manufactured in a laboratory by DuPont scientists doing molecular research into the theoretical base of polymerization, was quickly commercialized and marketed as a replacement for silk, specifically in the utilization as hosiery (cf. Meikle 1995: 125–52). The US American public were able to purchase nylon hosiery by May 1940. At the time, silk was being imported through Japan, and particular trans-Pacific tensions, which would reach critical mass later with the bombing of Pearl Harbour, were already straining this source. DuPont, further, was experiencing corporate image problems stemming from the politics of war in the United States, where DuPont had been accused of wartime munitions profiteering. In nylon, an entanglement of politics and materials emerges: The use of nylon to replace silk imported via Japan returned to the same region in the

---

[22] Latour introduces the term 'invention-discovery' at the outset of *We Have Never Been Modern* (1993) with respect to brain peptides, however does not elucidate what this doubling might entail, nor does he use the term again in this work. I am emphasizing invention here, at the same time I want to stress that I do not wish to insist on a binary distinction to discovery: Terms can be used in distinction without being diametrically opposed. (It is possible to describe something as aquamarine in distinction to turquoise without suggesting that one or the other colour *cannot* be blue-green.) Max Liboiron, for example, suggests that we might think of 'discovering' rather as 'laboriously crafting' (Liboiron 2021: 51).

form of parachutes and clothing in the humid jungles of the Pacific theatre of the Second World War (cf. ibid.: 148).

Replacement narratives do not necessarily follow this pattern, whereby a scarcity of a specific material results in the development of a replacement material. Gay Hawkins's analysis of the emergence of plastic bottles to displace those made of glass would be a case in point. Hawkins accordingly stresses 'technical *co-evolution*' (Hawkins 2013: 54, emphasis in original) to account for the emergence of different types of plastic as not only predicated on the practices of the lab looking to solve a particular problem, but also considering the ways in which products shape *future* or *potential* applications. Pushing this idea even further, Hawkins suggests: 'The emergence of a new material takes place not simply due to things that are imposed on it, but via creative processes of invention in which the material becomes 'richer and richer in information' (ibid.), where this information is crucially 'not simply molecular, *but also economic and social*' (ibid.: 55, emphasis in original). The story of the 'Post-It' is another example for the haphazard stories of chemical engineering and its applications, as it was initially a failed attempt to create a strong adhesive, but emerged materially 'richer' with its application as a temporary marker (cf. Molotch 2003). Cling-wrap, which I discuss at length in Chapter 4, provides another example.

Invention, thus, stresses the 'coming into' of material relations, which also pertains to the creative processes in the development of cultural artefacts. As readers, for example, we do not 'discover' fictional worlds as much as 'invent' them as part of the creative process of reading. Similarly, as interpreters of cultural artefacts, the meanings of artworks or interventions are not 'given', but engendered through processes of observation or interaction. This 'coming into' of material relations also encompasses processes of creation: The artist, in this sense, is no longer a genius or conduit, but an agent acting within a network of relations: an inventor rather than discoverer.

Mary Louise Pratt uses the term 're-invention' in her discussion of Alexander von Humboldt's works on exploration to stress the mediated quality of his numerous publications (cf. Pratt 2003, especially 111–43). Specifically, she notes that 'Humboldt's representations were not inevitable' that is, 'that their contours were conditioned by a particular historical and ideological juncture, and by particular relations of power and privilege' (ibid.: 127). Humboldt's natural historical endeavours, in her account, arise not only from the contingencies of his societal role at a particular time but also from material infrastructure and personnel: 'They [Humboldt and Bonpland] relied entirely on the networks of villages, missions, outposts, haciendas, roadways and colonial labor systems to sustain themselves and their project' (ibid.). Humboldt's discursive 're-invention' is thus shown by Pratt to rely, rather heavily, on inventory.

Inventory has its etymological roots in the medieval Latin *inventorium*, literally 'a list of what is found' and draws on *invenire* 'come upon' or 'come into', the same etymological root as invent. Indeed, Mary Carruthers argues that 'having "inventory" is a requirement for "invention"[,]… one cannot create ("invent") without a memory store ("inventory") to invent from and with' (Carruthers 2000: 12). The contemporary association of the word 'inventory' with the semantic field of business, as a complete list of goods, including, for example, property, stock or contents, also links to (violent) appropriation, profit maximization and risk externalization as capitalist practices.

Throughout this book, I note the ubiquity of plastic in our worlds. Plastic is undoubtedly inventory, pervading all aspects of our day-to-day life under a number of denominations, some more familiar than others: polyvinyl chloride, polyester, polyethylenes (as terephthalate, as high-density, as low-density), polyvinylidene chloride, polypropylene, polystyrenes (also as high impact polystyrene), polyamides, acrylonitrile butadiene styrene, polycarbonate and polyurethanes. Some objects and applications use more than one of these different kinds. Polyester and polystyrene are possibly the most familiar of this list, used in fabric and textiles, as packaging foam, disposable crockery and cutlery, and food packaging, amongst other uses. Other plastics are more familiar in their abbreviated forms: polyvinyl chloride as PVC, which we might encounter in shower curtains and plumbing pipes and guttering, polyethylene terephthalate as PET in drink bottles, plastic films and fleece, perhaps also polypropylene (PP) in bottle caps, appliances and car bumpers or polyethylene (PE) in supermarket bags and plastic bottles. Polyamides, better known as nylons, are found in fibres, toothbrush bristles and fishing line.

Other plastics are probably only familiar to most of us in the forms and objects they take, rather than the chemical composition names: In compact discs (this ages me), eyeglasses and traffic lights there is polycarbonate (PC); polyurethanes (PU) can be found in foams and coatings and are common in cars; polyvinyledine chloride (PVDC) in doll hair, cleaning cloths and artificial turf; and acrylonitrile butadiene styrene (ABS) is used in casings for electronic devices, small kitchen devices, Lego and other building blocks, as well as being inserted into the dermis in coloured tattoo inks.

And then there is an array of special purpose plastics or high-performance plastics: PTFE (polytetrafluoroethylene, the non-stick stuff of a non-stick frying pan – to be avoided, if you can), polymethyl methacrylate (PMMA for contact lenses, glazing (aka Perspex or Plexiglass)), phenol formaldehydes (PF, a thermosetting plastic used in Formica or insulating electrical fixtures), silicone (used as a sealant or in cooking utensils, and not to be confused with silicon, the element), polyetheretherketone (PEEK, a fantastic acronym, used often in medical implants, also in aerospace mouldings) and melamine formaldehyde (MF, used for break-resistant crockery), to name just a few. This exhausting, but not exhaustive, list of plastic forms gives rise to a number of crucial points: Plastic is omnipresent and takes many forms, that is, plastic (noun) is plastic (adjective).

Plastic's ubiquity, then, becomes accessible not through the myriad of polymers and chemical equations, but through the specific forms it takes. This is partly due to the secrecy surrounding the production of plastics.[23] It is also due to the sheer range of formulas used in the production (complicating plastic-abundance mitigatory efforts such as recycling) and, increasingly, in its 'micro-ubiquity' as micro-plastics.

The day-to-day experiences of plastic are tangible in a way that, for me, a non-chemist, the chains of polymers are not. I think, right now, of the plastic on the keyboard of my laptop, the plastic in my eyeglasses, the plastic components in my clothing, with which I am in direct contact as I write this, as well as the plastic insulation of the window frames, the various electronic auxiliary gadgets (scanner, printer), the sheen on some of the books in the shelves, the knobs on the central heating, the pens on the desk … . Plastic pervades my life. Like the many writers before me entertaining meditations on the extent to which plastic is ubiquitous, it seems that the idea is that such lists, such inventories, are formed to invite readers to likewise recognize the extent to which their lives are pervaded, permeated, by plastic. Chains of polymers give way to chains of objects.

Concrete also shapes my presence as I write these words. Entertaining the trope of a concrete inventory (as a riff on the more common all-the-plastic-things list, or concatenation of plastic; see Chapter 2), I can acknowledge a broad range of infrastructures and structures made of concrete that sustain and structure my life as I write these words: The concrete of the library building (both hidden, i.e. beneath the carpet and in the unseen basement and stacks, and rendered visible in a brutalist sense on the pillars that 'hold up' the structure, where the traces of

---

[23] The specific chemical compositions of the various materials encompassed by the term 'plastic' continue to be carefully guarded trade secrets; Heather Davis even writes of the 'nontraceability of plastic (which is partially due to the proprietary secrets of chemical companies)' (Davis 2022: 48). The inaccessibility of plastic production plants is a trope common to many documentaries on plastic, for instance, Werner Boote's *Plastic Planet* and Ian Connacher's *Addicted to Plastic*.

wooden buttresses remain visible and tangible). The concrete of the steps, roads, bicycle paths and gullies, the bridges passed under and over, the underpasses and tunnels on my way here. The concrete of the dams and power plants, and the bases for the steel pylons holding up electrical wires that feed my computer with its electricity. The concrete of the sleepers under the tram and train tracks, the platforms, bridges, tunnels, which all avail my movement when self-propelled transport won't suffice, and the runways and airports when the distances extend further. The concrete of sewage plants, riverbanks and the pipes through which the water passes that I take for granted when I turn on the tap. The concrete of the high-rises, sheds and bunkers that hold my data and emails …

Less immediate than plastic, perhaps, but no less crucial to any number of activities that I conduct on a daily basis with next to no thought for their reach: Concrete structures my life. Where lists of plastic tend to be quite concrete, enumerating any array of objects to evoke an accumulation of plastic, and my concrete list, with its radiating presences, rather more plastic. A sense of the *plasticoncrete*, again.

'Invention', and its close cognate inventory, is suggestive of imagined relations with materials. Evoking the inventory in this conjunction is to suggest that the 'objects', 'things' and, importantly, materials of invention are already present in the world, and that it is rather the *stories* of how these are integrated into history and other systems of knowledge that produce effects of modernity.

Modern materials: Manufacturing modernity

Concrete and plastic are constitutive of our modern lives. They have emerged with modernity *and* they shape modernity: They depend on modern structures for manufacture, and they form specific dependencies. Plastic and concrete, in the versions we are familiar with, both emerge in the nascent phase of European/Western Modernism. Histories of modern uses of concrete (more specifically: the invention of Portland cement) date to the middle of the nineteenth century, as do histories of pre-petroleum plastic (e.g. Parkesine, patented in 1856), each increasingly coming into their own throughout the twentieth century. They depend on the structures of industrialization for their production and dispersal through networks we know as 'logistics' for the extent in which they are pervasive across the globe today. For the moment, however, they are modern.

For Forty, concrete belongs firmly in a list of further technological practices (including antibiotics and the internal combustion engine) of modernity. He is explicit about this: Forty begins the first chapter of *Concrete and Culture* by asserting that 'concrete is *modern*' (Forty 2012: 14, emphasis in original). For Harkness, Simonetti and Winter, concrete is 'fluid and foundational, molten and rigid, moldable and structural, ancient and modern' (Harkness, Simonetti and Winter 2015: 309). They harness concrete's, or rather cement's, volcanic origins to interrupt its straightforward relation to modernity, offering *progressive modernity* instead, which is characterized by 'acceleration, speed and increased strength' (ibid.: 312) and as 'greedy and appear[ing] unbounded' (ibid.: 313). The widespread use of concrete in modernist architecture – Le Corbusier's works, for instance – works also to 'settle' concrete within the cultural semiotics of modernism and modernity.

Plastic, too, is intricately connected to modernisms and modernities. In Jeffrey L. Meikle's account, this has something to do with the simultaneity of economic, production and aesthetic factors coming together in the United States at a particular time. Indeed, Meikle argues that plastic is the 'essence … of modernity' (Meikle 1995: 107) – together with aluminium and stainless steel. This is not restricted to the United States, though. For Jody A. Roberts, the modern entanglement of human lives with plastic is one that reaches across the globe: 'Nearly every human subject found anywhere on the globe will likely bear the marks of a plastic modernity' (Roberts 2013: 128). With progressing research into microplastics, this is becoming even more true, as the planet's waterways are demonstrated to be increasingly permeated with plastic.

To suggest that the ways in which our modern lives are shaped by concrete and plastic is symptomatic of modernity is, crucially, not to suggest that modernity had to somehow already occur for these materials to shape our lives in the way that they do, nor, conversely, that modernity relies on plastic and concrete to come into being. Thinking materials, specifically concrete and plastic, means pushing back against such reductive equations of causality. Correlation is not causation. The relationship between modern materials (especially concrete and plastic) and modernity is processual and contingent, like the materials themselves: Again, they *are* shaped, but also *do* shaping.

To write so absolutely of concrete and plastic as materials of modernity harbours a specific set of problems, approximated by two questions: Firstly, what is modernity? Modernity refers, most simply put, to the current era. From the Latin *modo* 'just now', its deictic reference is to an ongoing period. Considering concrete and plastic as modern materials suggests attending to the ways in which they generate (and trouble) discourses of modernity, discourses which are often constructed along particular trajectories, both temporal and spatial.

If you ask 'what is modernity', you really need to be also asking 'and whose modernity is it?' If we accept that modernity in its Western mode emerged in the sixteenth century, thus contemporaneous to the discovery and settlement of the New World (Americas), then it was always relational to this new imagining of the globe.[24] There is another set of problems, closely related, that speak to the supposed *cultural* dimension of modernity – not quite the same as modernism, an aesthetic movement, but close – and the idea of an accultural modernity, a *modernization*. Unsurprisingly, postcolonial thinkers of modernity help to untangle this mess.

Let's begin with Gurminder Bhambra, who holds that 'colonization was not simply an outcome of modernity, or shaped by modernity, but rather, modernity itself developed out of colonial encounters' (Bhambra 2007: 77). With this, we come to recognize the ways Western *cultural* dominance was fed into, and corroborated by, an *accultural* view of modernity coded as historical progress. This has specific geopolitical, or at least spatialized, repercussions.

'Modernity', as Lars Eckstein and Anja Schwarz argue,

> is typically attributed to a momentous transformation *within* European societies following the conceit of 'rupture and difference' [the phrase is Bhambra's], a conceit that not only silences historical and transcultural entanglements, but also underscores a teleology of modernization according to a diffusionist logic which sets Europe at the global centre from where modernity then gradually spreads out across the remainder of the planet via the joint trajectories of colonization, mission and trade. (Eckstein and Schwarz 2014: 8, emphasis in original)

This mode of thinking about modernity, sometimes employing the plural form, that is, 'modernities', has the effect of shifting away from an 'inevitability' and positivity (or 'universalism') of modernity as a *singular* path to progress. It also pushes back against the notion of a nexus from which modernity somehow 'emerges', either radiating out of (spatially, from some kind of centre) or following on from (temporally, from some kind of point of origin). It further

---

[24] Note, at the same time, that to focus on the expansion into the 'New World' before, during and after the sixteenth century should not entail turning a blind eye to colonial patternings that were taking place and expanding throughout other regions of the world. The moment of contact might attract a lot of attention, particularly in the vein of the 'Columbian exchange' of plant and animal species, as well as disease and in more explicit practices of genocide. The patternings for European developments occurring in the Americas (here, 'modernity') were prefigured in many ways by contact and trade between Europe and Asia as well as Europe and Africa, and, as Patel and Moore argue, were tested at more regional levels, is in their case study of Madeira (Patel and Moore: 2020).

works to destabilize the trajectories of modernity that pretend to the 'discovery' of modernity in the 'West'.[25]

To consider concrete and plastic as modern materials is to realize that they are of 'the current age', which in turn leads to considerations of the Anthropocene (see the section 'Condensing relations: Materialities in the Anthropocene'). If we conceive of transformations of materials and material practices as invention, as a 'coming into relations', the nexus of modernity discourse shifts from 'point of origin' ('a central or focal point', 'Nexus') to 'patterns of connection' ('a network', 'Nexus'). Modernity, as it pertains to materiality, no longer relies on discourses of originary-ness.

So, modernity is both condition for and conditioned by those materials we might consider modern. Materials, as Elizabeth Shove et al. argue,

> do not exist outside society. They are not 'just there' waiting to be exploited, discovered and appropriated. In materials science, as in other areas of scientific endeavour, lines of enquiry and pathways of innovation are nudged this way and that by socially and historically specific patterns of investment, interest and enthusiasm. The properties and characteristics of man-made [*sic*] materials consequently reflect and embody characteristics both of the culture in which they were made and of an imagined future in which they might be used. (Shove et al. 2007: 97)

Material modernity patterns through any number of material accumulations and practices. Considering concrete and plastic as modern is also to recognize materials as relations in the processes of production, invention, consumption, disposal, transformation, decomposition, adaptation and so on. Another way to phrase this is to propose that the relations of concrete and plastic to modernity are entangled. The term is suggestive of the materiality of threads: I think of the capacity of headphone cables, and necklaces, fibre threads and other things, to become knotted seemingly without any readily identifiable agency (that is, without me tying them into knots).

Entanglements, writes Ian Hodder, 'are held taut by dependences, affordances, abstractions, regulations and resonances while at the same time being open,

---

[25] The 'West' was coded above as 'European societies', and in Forty and elsewhere a similar construct called the 'North' or 'Global North' comes up; all of these shorthand phrases harbour their own inconsistencies. Max Liboiron, writing of science practices, is able to sidestep the problematic spatialization of such divisions (West/rest; North/South) by proposing 'dominant'. The qualifier dominant, they argue, insists on power relations and allows for grappling with absolutist or universalizing tendencies as well as leaving room for non-dominant approaches within the spatialized spheres (cf. Liboiron 2021: 20–1, fn 77).

complex, unruly and contingent' (Hodder 2012: 176). Key to Hodder's notion of entanglement is the mobilization of the various meanings behind 'depend': the two main meanings suggest, on the one hand, the ability to rely on, require or trust; on the other hand, 'depend' suggests controlling or limiting circumstances. The entanglements of humans and matter through dependencies do not follow a trajectory of intent, but, as Hodder notes, 'Change of entanglements tends to be directional in that it is difficult to reverse human-thing dependences' (ibid.: 177), a point that I have alluded to above in thinking through invention and inventory. This increase in entanglement is not, Hodder stresses, teleological (cf. ibid.: 167), although 'rates of cultural change have increased as the scale and complexity of human-thing entanglement have increased' (ibid.: 177). Modernity, read against these observations, becomes a mess of problems that have been solved in different ways, or, that have only emerged together with solutions to other problems.

Concrete and plastic have been and continue to be used as articulations of modernity. The particular practices of engaging with concrete and plastic do not follow a prescribed logic, do not radiate from some kind of centre, nor do they 'speak' to some kind of origin. We are, nevertheless, entangled deeply in the material practices engendered with, through or, following Eve Kosofsky Sedgwick, 'beside',[26] modernity. Modernity, drawing on an accumulation and an entanglement of material practices, in turn results in the accumulation of materials, as becomes most evident in the next section, where I trace another way of thinking about our current age: the Anthropocene.

Condensing relations: Materialities in the Anthropocene

The idea of condensing relations draws on the materiality of materials to work. It suggests the ways in which the dimensions of materials extend, or stretch, and condense through practices, forging relations over (and through) time and

---

[26] Sedgwick argues:

> '*Beside* is an interesting preposition also because there's nothing very dualistic about it; a number of elements may lie alongside one another, though not an infinity of them. *Beside* permits a spacious agnosticism about several of the linear logics that enforce dualistic thinking: noncontradiction or the law of the excluded middle, cause versus effect, subject versus object. Its interest does not, however, depend on a fantasy of metonymically egalitarian or even pacific relations, as any child knows who's shared a bed with siblings. *Beside* comprises a wide range of desiring, identifying, representing, repelling, paralleling, differentiating, rivaling, leaning, twisting, mimicking, withdrawing, attracting, aggressing, warping, and other relations. (Sedgwick 2003: 8, emphases in original)

space. It harnesses the globalizing and localizing tendencies of concrete and plastic to address, most specifically (but by no means exclusively) anthropogenic alterations to the environment. Condensing works both spatially and temporally, although the later will be more prominent in this section.

Condense, as a verb, has two meanings. As a transitive verb (taking an 'object') it entails the grammatical object becoming more concentrated, as in a soup becoming thicker, or when an object becomes more concise, as in *Reader's Digests* (where the assumption is that the idea of the story remains, it's just expressed with less words). As an intransitive verb, 'condense' suggests the changing of states, specifically, from gas or vapour into a liquid. I take some liberty to harness this second meaning in suggesting condense might also mean the ways in which substances become more *fluid*, or indeed less.[27] The flows of materiality are the concern of this section, which traces the ways that materiality stretches across the globe, and condenses, thickens, intensifies, magnifies, accumulates, coagulates, accretes or erupts in particular ways.

I choose condense specifically to elicit these meanings of thickening, becoming more concise, and changing states. Close cognate 'folding' suggests a 'hiding' into the folds and does not have the tangible thickening immediately, only as a consequence of repetitive folds, as in folding a piece of paper. 'Emerging' is another approximation, but it evokes the visual paradigm of light and knowledge and does not draw out the dormant or remnant layerings of the material as in condense. Condensing, then, recollects the small shifts in materialities of increased intensity, but reckons with the flows of materiality that stretch to encompass the globe.

Concrete and plastic are materials of industrial production. Both materials are subject to global networks of extraction (e.g. mining), manufacture (e.g. in factories) and dispersal (e.g. logistics, also of disposal) throughout the globe. All stages require labour and are extractive. To recognize the global reach of these materials is, crucially, not to suggest the evenness of the spread of these materials. Approaching materialities entails reckoning with their uneven relations. As plastic and concrete shift from materials to objects through process of production and pouring or moulding, the ways in which they entangle our lives at various stages of these processes are not so straightforward.

[27] Although sometimes used interchangeably, *fluid* refers to a substance that 'has no fixed shape and yields easily to external pressure; a gas or (especially) a liquid' ('Fluid' 2011) from Latin *fluere* 'to flow', whereas *liquid* refers to a substance that 'flows freely but is of constant volume, having a consistency like that of water or oil' ('Liquid' 2011) from Latin *liquere* 'be liquid'. Fluid is the broader term, covering liquids, gases and plasmas. Taken strictly, condensing only applies to liquids, hence some 'poetic' or theoretical license is mobilized here.

Concrete is both localized and globalized. Concrete harnesses local components – the aggregate and the water – as well as global components – cement, and, on occasion, steel. Cement, its manufactured component, has a definitive global dimension: the 'most abundant of all manufactured solid materials' according to van Oss, with a yearly output estimated at 2.5 metric tons per year per person on the planet (van Oss 2005: 2, it has increased in the meantime, cf. VDZ 2018).[28] As Forty suggests, 'Few, if any, other technologies have made the transition from West to East, from North to South, quite so completely as concrete' where, he claims, 'It is the generic medium of construction, a means to an end – modernization – used with unselfconscious abandon' (Forty 2012: 119). If concrete is grounded, if it is *concrete* in this sense, then it is inherently also manifest in the shifting flows of globalized production, logistics and demand.

Concrete stretches globally and yet *condenses* at particular sites. One concrete × concrete example is offered by the structures of reinforced concrete. This not only concentrates the materials of concrete in particular locations – where the masses of concrete give way to protuberances of high-rises extending both visibly into the air, but also hidden under the earth in the bases that sustain these heights.[29] Here, the concentration of concrete also concentrates relations, rather obviously also due to the greater density of humans, and nonhumans, per square metre, as the plane of the earth is multiplied through the storeys of the high-rise (see also Chapter 5).

Plastic presents a similar pattern of stretching and condensing. Rates of the world production of plastic continues to increase yearly ('Plastics Europe'), meaning not only continually more plastic, but *increasingly* more plastic.[30] The

---

[28] This statistic, based on 2004 figures, corresponds to roughly 1.7 cubic metres per person/year of cement. The output of concrete is much larger. A cursory search on the internet to update the statistics suggests that the proportions of cement:sand:aggregate are approximately 1:3:3 (in weight), and that cement production continues to increase (VDZ 2018). Let me just say: there is a lot of concrete being made all the time.

[29] Such structures give rise to that particular icon that entangles 'city' with 'globalization' and 'capital': the skyline. Capitalism does not, in any way, have exclusive proprietorship of concrete high rises. The use of concrete for the construction of dwellings was widespread in the Soviet states, most prominently in five-storey types and, later, in sixteen-storey types. Nor are the high rises exclusive to the 'West': look to (South)East Asia, look to the Persian Gulf, and behold, there they are, (very, very) tall, concrete buildings, making very specific entanglements of globalization/modernity/capital (very, very) explicit.

[30] Graphs of growth rates obscure the way in which matter accrues. Looking at the statistics provided by PlasticsEurope and Consultic ('Plastics Europe'), the figures for the years 2008 to 2016 rise from 245 million metric tons to 335 million metric tons *per year*. This reflects increases of 90 million metric tons/year over eight years. But the weight of new plastics produced in the eight-year period is not 90 million metric tons, as a cursory look at the graph might suggest, but 2,599 million metric tons. We need to exercise caution in reading graphs: some actually show change over time, not accumulated difference. Max Liboiron notes that 'one estimate states that half of all plastics made throughout history has been produced between 2004 and 2017, and while this figure is based as much on charisma as math, it does illustrate exponential growth' (Liboiron 2021: 96).

global reach of plastics – encompassing the omnipresence of microplastics,[31] the ubiquity of bisphenol A and the plastic bag – condenses in specific forms and in specific relations: in a whale replica, for instance, made entirely of plastic, stranded on a beach in Naic near Manila (Philippines), replete with colourful plastic objects in its mouth. Such images condense the issues of plastic into statement images about wasteful plastic practices. If plastic accumulates – the etymological roots suggest a 'heaping up' – this is an image of piles of objects rendered waste.

My preferred term, condense, instead recollects the intensity of material increasing, as well as the engagements (and agencies) of it; references to the 'stretchiness' of patternings also allude to the plasticity of plastic. Indeed, the long durability of concrete and plastic binds resources – raw materials – in particular forms. The materials of plastic and concrete draw on raw materials – petrochemicals (or corn and other monocultural crops for bioplastics), limestone, sand and stones – deep time 'condensed' into deposit. For, concrete and plastic derive from fossilized or calcified organic life, found throughout the globe.

These entanglements also extend into the recent fossil record: 'Recent anthropogenic deposits contain new minerals and rock types, reflecting rapid global dissemination of novel materials including elemental aluminum, concrete, and plastics that form abundant, rapidly evolving "technofossils"' (Waters et al. 2016: 137). The resources of the earth are finite. Thinking with condensing (and also the 'stretchiness' of these materials), we can attend to the way that new materialities are generated. And with them, new forms of interactions, of practices and of relations. Recycling materials is, typically, reusing materials and regenerating materialities: It is most emphatically *not* 'returning resources'.[32] As Gay Hawkins, Emily Potter and Kane Race observe of PET bottles, 'The "raw material" is not easily accessed because it is structurally integrated into the bottle, which is designed to be indestructible' (Hawkins, Potter and Race 2015: 132). Whilst recycling, on the surface, suggests a 're-turn', it is not a simplification of relations but rather a magnification of entanglements.

---

[31] Eric van Sebille et al. note that plastic debris has been documented in *all* marine environments, 'from coastlines to the open ocean …, from the sea surface to the sea floor …, in deep-sea sediments … and even in Arctic sea ice' (2015). The global reach of plastics is also evident in the 'virtually ubiquitous microplastic particles (microbeads, "nurdles", and fibers)', as Waters et al. argue in *Science* (2016: 140).

[32] The impossibility of this 'return' is given fictional form in Kurt Vonnegut Jr.'s *Slaughterhouse-Five*, a passage that has stayed with me for many years. I turn to this passage in the conclusion in more detail.

This book directs attention towards the materiality of these entanglements, and their condensations, and hence fits within the remits of discourse on the Anthropocene. The proposal to introduce the Anthropocene as a geological epoch draws attention to the presence of a geological strata dominated by recent human origin and is also evoked by the cover of this book, depicting a mosaic called 'We Were Here Now' by artist Julie Sperling. Like modernity, the Anthropocene enacts a double gesture: It appears to continue on (we remain within it) even as it points away from itself for its referent point (to the past, from which we are continually distancing ourselves), continually slipping away from an imagined origin.

The sense of the unevenness inherent in the *anthropos* of the Anthropocene is pervasive within postcolonial, decolonial and critical race studies' critique of the Anthropocene.[33] This critique is particularly, but not exclusively, articulated as pertaining to responsibility, in both the sense pertaining to 'fault' and the sense of 'dealing with it'. Dipesh Chakrabarty, for instance, asks: 'Why should one include the poor of the world – whose carbon footprint is small anyway – by use of such all-inclusive terms as *species* or *mankind* when the blame for the current crisis should be squarely laid at the door of the rich nations in the first place and of the richer classes in the poorer ones?' (Chakrabarty 2009: 216). Other voices critique the purported evenness of the *anthropos* by addressing other aspects, for instance, Audra Mitchell, who stresses the 'links between the forms of agency, power and violence that have contributed to the Anthropocene' (2015). For the Pacific Islands, as Peter Vitousek and Oliver Chadwick point out, the temporalities of the Anthropocene shift in various ways: Among the 'last places reached by humanity', some of the islands have 'entered the Anthropocene early', despite 'no access to energy from the past' or 'borrowing from the future' (Vitousek and Chadwick 2013). For many, the effects of the Anthropocene are a present presence, not a future threat.

Debates of the Anthropocene led in the field of humanities are concerned with the *discourses* of the Anthropocene. As the Anthropocene discourse has its genesis in geology, it is also preoccupied with markers, that is, with material manifestations of human incursions into the world. Importantly, the proposed markers for the Anthropocene are indexical: The relations through which it

---

[33] See, for example, Parikka (2014), Whyte (2017), Nixon (2018), Davis et al. (2019) and Armiero (2021). As I have argued elsewhere (cf. Crane 2019b), the crisis of the 'crisis of imagination' rests not with the unwillingness of artists and authors to engage with the Anthropocene (or Climate Change), but rather with the incapacity of certain readers to bridge the rift between their own (privileged) lives and the (depictions of) lives which must confront the effects of the Anthropocene already.

enters discourse are not reliant on metaphors or other connecting structures that must, somehow, bridge gaps. Indices are material evidence, albeit nominated by specific bodies in response to particular discourses. The date invoked as the *start* of the Anthropocene is hence more than a jostling for discursive power: The Anthropocene, by evoking long geological timescales, asks us to look back at what has come before and towards what will come (cf. also Colebrook 2014: 24).

Accordingly, if 1610 is evoked as a start date, and the *drop* of carbon dioxide levels is proposed as a marker, it is about a story of the Anthropocene that coincides with European colonization of the American continent (cf. e.g. Lewis and Maslin 2015).[34] This is a powerful story in that it asserts the unevenness of that unsettling prefix 'anthropo-', pushing back against its insinuated universality. It is an unsettling, an uncomfortable shift or violent dispossession, which finds articulation in many other alternative names for the Anthropocene, such as the Plantationocene (e.g. Davis et al. 2019), the Capitalocene (e.g. Moore 2017, 2018), the Anthrobscene (Parikka 2014), the Chthulucene (Haraway 2016) or the Wasteocene (Armiero 2021).

If 1945 is chosen as a start date for the Anthropocene, or more specifically 16 July 1945, then it is clear that another kind of marker is put forward: radionuclides. The date in July is accompanied by a specific site, Alamogordo, New Mexico, USA, and corresponds to the first detonation of an atomic device. Elizabeth DeLoughrey's evocation of the 'Nuclear Pacific' (cf. 2012) and, in another genre entirely, Isao Hashimoto's animation of the nuclear explosions from 1945 to 1998 (see video available, for example, at youtube.com), speak to this frame and its ongoing and disastrous effects.

Various other markers have also been proposed, suggesting different processes and foregrounding other interactions with the environment.[35] Of particular pertinence to this book are those that pertain to the materials of plastic and concrete. One marker is particularly pertinent, 'plastiglomerate', which is 'an indurated, multi-composite material made hard by agglutination of rock and molten plastic' (Corcoran, Moore and Jasvac 2014: 5): a *plasticoncrete* artefact.

---

[34] The dip in carbon dioxide levels proposed suggests the incursions of a particular modernity upon other cultures and hence has purchase in postcolonial-informed critiques of the Anthropocene. However, Waters et al. suggest, following Jan Zalasiewicz et al.'s article 'Colonization of the Americas', 'these fluctuations do not exceed natural Holocene variability' (Waters et al. 2016: web). Further, the counter-intuitiveness of the marker – a *reduction* in CO_2 levels – works against the narrative force of GSSPs as cautionary tales.

[35] I choose the term 'unsettling' with care: It works in conjunction with the patterns of ('historical' and continuing) settler colonialism suggested by this starting date.

Corcoran, Moore and Jasvac suggest that plastiglomerate is suited as a GSSP for three key, interrelated, reasons: First, in contradistinction to some other proposed markers, plastiglomerate is in itself an anthropogenic material. It is an index, a kind of 'sensory feature' which links or correlates the sign with what it represents rather than an abstract sign. It does not indicate some other practice, but is itself the practice for which it also acts as marker: the manufacture of plastic, as evidenced by plastic. Second, plastiglomerate should be available throughout the planet (the authors suggest that 'the global extent of plastic debris could lead to similar deposits where lava flows, forest fires, and extreme temperatures occur'; Corcoran, Moore and Jasvac 2014: 7). Third, it is mostly non-degradable. Although these characteristics are usually considered amongst the greatest environmental problems associated with plastic, the distribution of plastic across the planet (both land- and ocean-based), including those places where conditions are felicitous for the production of plastiglomerate (i.e. concentrated heat and other rock matter), as well as plastics 'longevity', all make it an apt marker – an index – for the Anthropocene.

In plastiglomerate, *plasticoncrete* comes into its own: The cement of concrete is replaced by plastic, which acts as the binding substance between various aggregates. Industrial processes are mimicked (it is not, after all, concrete); materialities gain agency (the production of plastiglomerate does not spring from human intention); it speaks to larger material processes and practices (and their spatial spread) and it condenses relations. Its capacity to act as a marker for the Anthropocene is suggestive of its stretching towards an unknown future; its petrochemical source stretches into a prehistoric past.

However, taking plastiglomerate as a marker harbours the danger of reducing the complexity of human entanglements in their environments to a single issue, even as this choice of marker shifts agency and responsibility to the large number of humans who partake of plastic practices. At the same time, changes in sea levels or temperature run the risk of reducing the complexity of human-induced environmental changes, for which 'Anthropocenes' act as a theorized shorthand, to a very specific set of indices: Climate change continues to dominant discussions of environmental crises, at the risk of marginalizing other issues. Contextualization, interpretation and narrativization are therefore crucial. These practices – contextualization, interpretation and narrativization – comprise the place of the (environmental) humanities in the discussions of the Anthropocenes.

My enthusiasm for plastiglomerate is, rather obviously, a function of the foci of this book: It has not seen widespread adaptation.[36] Waters et al.'s 'The

---

[36] The International Commission on Stratigraphy's Working Group on the Anthropocene has suggested the mid-twentieth century – the beginning of the nuclear age and the Great Acceleration

Anthropocene is Functionally and Stratigraphically Distinct from the Holocene' offers a succinct and comprehensive discussion of the debates. Ultimately, they refrain from nominating one particular marker for the commencement of the Anthropocene. They note that the 'implications of formalizing the Anthropocene reach well beyond the geological community' (Waters et al. 2016). Their open conclusion – with its insistence on the Anthropocene as resulting from the activities of, and simultaneously being witnessed by, 'advanced human societies' (ibid.) – is powerful in that it asserts responsibility as well as, through its very inconclusiveness, acknowledging a set of conditions that provide for the simultaneity of several narratives for the Anthropocene.

The multiplicity of narratives allows for a multiplicity of actions in a multiplicity of time frames, that is, in the past, present and future, and spreading unevenly across humans and non-human agencies. The global scale of the Anthropocene era should not be taken to suggest an evenness of Anthropocenes. Contemporaneous human actors (as one set of entities that occupy the earth) do not experience, contribute to or feel the effects of, 'the' Anthropocene in the same ways. Rob Nixon accordingly suggests: 'We may all be in the Anthropocene but we're not all in it in the same way' (Nixon 2018: 8). We need many stories to account for this, as Waters et al. suggest: We will need to become 'inventive', to come into relations, with (unknown) others, stretching across space and time.

Future artefact

I use the term 'future artefact' to stress that manifestations of material present (both temporally and spatially) imply larger temporal and spatial dimensions.[37]

in greenhouse gas emissions – as the official start of the Anthropocene, stressing the presence of artificial radionuclides as *one* material marker. As Zalasiewicz et al. explain, 'The artificial radionuclides scattered around the Earth may be regarded as a primary, and arguably the primary, marker for Anthropocene strata because of their global distribution, relatively easy detectability and near-synchroneity of expression, which broadly coincides with multiple signals of significant environmental change during the mid-20th century' (Zalasiewicz et al. 2017: 86). They also note how this particular marker might appear 'environmentally trivial' in comparison with the scale of other 'anthropogenic perturbation[s]', even considering the devastating explosions that saw the end of the Second World War in Japan (Hiroshima and Nagasaki), and the ongoing fallout across the planet (cf. also Masco 2021). In the article, by members of the International Commission on Stratigraphy Working Group on the Anthropocene, the authors also suggest 'turning' the 'approach on its head' to find 'environmental trends picked out as of major significance to contemporary global change by the ESS community and consider whether or not they will leave a recognizable signal within strata that may then be used as a basis to create chronostratigraphical units' (ibid.: 87): science being speculative.

[37] The future artefact also recognizes another reason for the obsession with the markers for the Anthropocene, the 'golden spike' or Global Boundary Stratotype Section and Point (GSSP): Impact. As Raj Patel and Jason W. Moore pithily note: 'Future intelligent life will know we were here because

It is an instance of the 'stretchiness' of materiality condensed into a material record. The two terms – future and artefact – work together to reference a deictic deferral in the speculative mode, with material effects.

The 'future' of the future artefact is an imaginative heuristic that identifies as 'future' the time to come after realization or creation (i.e. either as material artefact or as a representation thereof). The 'future' of the artefact might *already be* the present and presence in another spatial dimension. I think here of the deferrals of globalization and waste. Such an enveloping of temporal dimensions is imagined by Karen Barad as a process of sedimentation: 'The past matters and so does the future', they note, 'but the past is never left behind, never finished once and for all, and the future is not will come to be in an unfolding of the present moment; rather the past and the future are enfolded participants in matter's iterative becoming' (Barad 2007: 181). The 'future artefact', with its folding of (geological) times, and, importantly, with its residues, allows for the materials that shape our lives to be imagined shaping others: 'elsewhere' and, with some creative license, 'elsewhen'.

Following Claire Colebrook, a sense of (temporal) deferral is built in to the Anthropocene, which 'relies on looking at our own world and imagining it as it will be when it has become the past' (Colebrook 2014: 24), which will, as distinctive geological strata, become 'readable' (ibid.) in the future. In other words, the Anthropocene, to become a 'proper' geological era, requires a marker of the *past* that imagines a *future*, which then becomes a heuristic for understanding the *present*. This analysis finds the Anthropocene 'present' folded into temporal dimensions of the past and the future; these are 'sticky folds', like the stickiness of cling wrap (see Chapter 4). It presents entanglements and dependencies. In cultural texts – fictional or non-fictional, narrative, poetic, audiovisual, monumental, as well as quotidian materials – these practices of becoming 'readable' give rise to interpretations of the (Anthropocene) present.

This 'present' suggests a temporal dimension. However, this 'present' also has a spatial dimension. As scholars working to decolonize the Anthropocene argue, the 'other' to which the Anthropocene points is not exclusively an 'elsewhen' but also an 'elsewhere'. Kyle Powys Whyte uses the phrase 'today's dystopia of our ancestors' (Whyte 2017: 208) to make the point that the folding of temporal frameworks that sometimes accompanies the Anthropocene is not limited to

some humans have filled the fossil record with such marvels as radiation from atomic bombs, plastics from the oil industry, and chicken bones' (Patel and Moore 2020: 1). They also point at externalities interrupting the capitalist system (cf. ibid.: 21), which is a way of considering how material remnants of systematic exploitation cannot be so easily disappeared.

the future (i.e. what is to come from the present moment), but *also* as a way of thinking about the present from a different, non-dominant (temporal, spatial) standpoint. Relationships with 'plants, animals and ecosystems', which in the Anthropocene are altering at a 'wrongfully rapid place' Whyte argues, is *'another kind of anthropogenic environmental change: climate destabilization'* (ibid., emphasis added) that follows from the anthropogenic environmental change enacted in (settler) colonialism rather than a new phenomenon. The effects of the Anthropocene are *not* restricted to 'elsewhens', and, as the work of scholars such as Whyte demonstrates, 'elsewheres' are a function of considerable privilege. The tangible effects of the Anthropocene are not exclusive to an (imagined) future, but are very *present* in the quotidian lives of many people alive today, already living with rising sea levels, toxic waste, extreme weather events, mass extinction events and loss of habitat.

The etymology of 'artefact', in turn, has two components: *arte* 'by or using art', and *factum* as the past tense of *facere*, 'to make'. Both components signal processes of human intervention, as well as invention, and are crucial for this book with its interest in human-produced materials. Artefact is a noun developed out of an adverb and a verb, and, at an etymological level, harbours processes that result in material consequences. Whilst the emphasis of 'future' is to suggest the temporal (and spatial) dimensions that stretch far beyond the artefact in line with arguments rehearsed through the Anthropocene, the term 'artefact' reminds us of the techniques and technologies – the practices of 'making' – that engender such points of interest.

One meaning of 'artefact' pertains to an object, typically of historical or cultural interest. This is the component that most readily stretches towards the postulated audience of the future, some kind of 'future archaeologist' (I return to this later). The 'artefact' *also* references something residual in scientific experiments and manipulations, the kind that allows Karen Barad to write of the 'separation of fact from artifact' (Barad 2007: 35). The 'residue' component, whilst often backgrounded, rationalized (as a by-product of production, as waste, itself an inherently relational term) or ignored, not only mobilizes the environmental concerns of the Anthropocene, but also – crucially – stresses relations of materiality.

The 'future artefact' carries these insights with it, in its blurring of temporalities and spatialities, insisting on materialities and imaginations. Whilst the concept 'future artefact' is my own, that is, one that finds specific articulation within this book, it does have some close cognates.[38]

---

[38] In the slip of time between originally writing this introduction and doing the revisions – about four years in this case – I have come across even more related ideas. Several of these ideas require

I have already alluded to one of these: the 'future archaeologist'. Heather Davis evokes this figure in her deliberations on plastic: 'Read through the lens of the fossil', she suggests plastic 'conflates geology with archaeology, figuratively imagining a future archaeologist who will discover these fossils' (Davis 2022: 43). Nick Yablon traces the figure back to the *New York Times* asking its readership in 1925 'to imagine their city as it would appear to a "dark-skinned" [*sic*] archaeologist in some distant time when its skyscrapers have been eroded by rain, snow, ice, wind, and rust' (Yablon 2009: 1).[39] David Farrier also mobilizes a related figure, a 'future geologist' who will be able to trace shipping lanes through the sediment of clinker left behind by ships upon which, he postulates, a 'plastic seafloor trail' of jettisoned waste will be evident (Farrier 2020: 53). With the 'future artefact', the interpellative function of the artefacts bypasses their 'thinginess'.

In an analysis of J. G. Ballard's *Concrete Island*, Roger Luckhurst articulates another similar idea. With the phrase 'ruin of the future' (cf. Luckhurst 1997: 137) he conjures the idea of future decay contained in the present artefact. Luckhurst argues that 'the effect of the ruin works, as in [J. G. Ballard's] *Concrete Island*, to reveal the erasures of the phantasmagoric, simulacral world' (ibid.: 140). The 'ruin of the future' touches on some of the ideas of the 'future artefact', yet in referencing simulacra in this way, it emphasizes the functions of representations and mimesis and hence neglects the opportunity to engage more fully with materialities. Whilst the ruin is crucial to the interpretations of my final chapter, Luckhurst's formulation appears to resist the interpretation of ruin as verb (see Chapter 5). Further, the ruin does not seem to map onto the materialities of plastic as well as those of concrete; hence, I prefer 'future artefact'.

Two further concepts also do similar work to the 'future artefact': 'future remains', figuring most prominently in the publication of a volume by the same title (Mitman, Armiero and Emmett 2018),[40] and 'future fossil', which finds articulation in several locations, including Rob Nixon's *Slow Violence and the*

---

object integrity; however, the 'future artefact' does not necessarily require objects, but emphasizes materials. I have also had a lot of fun over this period teaching with the idea, including creating virtual exhibitions with students.

[39] Yablon remarks, that 'the *Times* remained uncertain whether this archaeologist ... could reconstruct the lost civilization from any of these surviving ruins and relics. The evidence might be too fragmentary to illuminate its technological, let alone cultural accomplishments' (Yablon 2009: 2). That this future archaeologist is 'dark-skinned' speaks to cultural fears of an altogether racist mode of anxiety.

[40] This volume comprises a revised and print form publication of considerations of the 'Cabinet of Curiosities' online project of the 'Anthropocene Slam', based out of the University of Wisconsin-Madison. The website is currently only accessible via cached copies: Its digital manifestation is now archived, itself an artefact of sorts.

Environmentalism of the Poor and, amongst other sites, Jussi Parikka's *A Geology of Media* and, prominently, in David Farrier's *Footprints: In Search of Future Fossils*.

In the preface to *Future Remains*, the editors argue: 'The orientation of history is up for grabs – as are the objects that make up history's archive, that foreshadow the future, and that will bear witness to a future past' (Mitman, Ariero and Emmett 2018: ix). The ways in which the objects invite speculation regarding what might *remain* in the future is the key impetus to their volume. However, the perhaps less obvious articulation of the second term – how the future *remains* to be of concern, where remains is mobilized as verb – is not explicitly referenced. They thus forgo the opportunity to dwell on the way 'future remains' suggests *both* that which will remain in the future *and*, less obviously, how the future remains a concern of the present. Most critically for my book, Mitman et al.'s emphasis on objects falls short of a comprehension of materiality in terms of its materials. When Mitman et al. highlight the jar of North Carolina sand proposed by Tomas Matza and Nicole Heller in their preface, their emphasis on it as object means they forgo the opportunity to articulate the way in which the jar itself is made of sand. The 'artefact', with its double meaning, allows for such considerations.

Jared Farmer's 'Future Fossil' alludes to the idea of the future artefact, even as it does not completely redeem its 'promise': The artefact itself is absent. The 'object' is, instead, the trace of the Blackberry. Farmer's use of 'Utah mudstone from [his] late father's rock collection' (Farmer 2018: 191) references the presences of dinosaurs in this state (Utah has an abundancy of dinosaur sites) and is combined with a clay of polymer 'a petroleum byproduct' (ibid.), which in turn correlates with the absent plastic object. The idea of the fossil suggests a living specimen, from which the traces extend: where Farmer suggests that his 'art piece satirizes the relationship … between consumer capitalism and biological extinction … extending the 1960s idea of "product life cycle"' (ibid.), its materiality goes further, it *stresses* such a relationship. Thinking of Farmer's future fossil as a form of future artefact thus recollects the second meaning of 'artefact', namely the residue (the unintended, surplus, supplemental component).

In noting 'two overlapping spheres' of the 'discourse of the Anthropocene', Farmer's contribution to *Future Remains* elicits the double valency of the 'remains' of the title, as 'stratigraphic record' that will 'exist millions of years from now' and as a 'rhetorical description of human influence on the complete Earth System … in the current moment and the recent past' (ibid.: 193). Interpreting

Farmer's 'Future Fossil' as 'future artefact' opens analysis to the materiality of the forms, both present and absent.

The phrase 'future fossil' also emerges in the works of Jussi Parikka, for example, in his *A Geology of Media* in the title of chapter 5 ('Fossil Futures'). Parikka 'speculat[es] on the idea of future fossils, as a future temporality turned back to the current moment. The fossil is in this sense a question about the contemporary that expands across multiple times ... [and a] material monument that signals a radical challenge to prevailing notions of time' (Parikka 2015: 109). This is particularly pertinent given the otherwise ephemeral conceptualizations of data and technology, rendered less tangible by metaphorical designations such as 'the cloud'. Even decentralized data storage practices have physical manifestations: Strings of 0s and 1s require matter to be stored, transmitted and received, as well as energy required for these processes, as Parikka is also careful to note.

Some artefacts do the future 'better' than others (which is to say that they are either more convincing or simply more captivating to the imaginations of their creators or interpreters). We might consider 'future ruins' – for example, Joseph Michael Gandy's imagination of the decay of the Bank of England over a period of 100 years, first exhibited in 1830 (cf. Parikka 2015: 117). We might also consider a snow globe, as in Emily St. John Mandel's *Station Eleven*, where folds in temporalities becomes increasingly evident through its inclusion in the curated exhibition entitled 'Museum of Civilisation'. The plastic snow globe – imagined passing through logistics of labour and exchange rendered defunct in the imagined future of the novel and yet undoubtedly present as an artefact – does this work, at the same time as the image of snow forges connections to other scenes of the novel, specifically the *plastic* snow on the stage at the outset of the novel.

The terms of 'future artefact' interact to elicit a temporal framework that reaches into not only what is 'to come' – the *venire* of invention, inventory – but also what 'has come'. The first valency of artefact, in the meaning that evokes culture and history, also stretches to the past, thus interrogating, and expanding on, future trajectories. Here, then, is a site through which the temporal stretch of materiality comes into its own, so to speak: evoking non-anthropocentric, primordial origins at the same time as non-anthropocentric, speculative fates.

On materiality and Material Culture Studies

Concrete and plastic are materials. They are, however, not really the kind of materials that 'Material Culture Studies' takes as its subject, or, to make a point of it, as its *object*. Material Culture Studies typically examines the physical manifestations of culture, and the practices and rituals that accompany these artefacts, and thus encompasses disciplinary fields such as anthropology, archaeology, art history, ethnography, history, museum studies and so on. These disciplines are to be found in the humanities and social sciences, rather than the natural sciences: Material Culture Studies is, by way of such a location, more interested in the workings and contextualizations of 'Culture' than in those of 'Material'. Andrew Pickering, in his critique of this divide, suggests that this is a remnant of Cartesian dualism: 'Matter (nature, machines, instruments) is the stuff of the natural sciences and engineering, but knowledge belongs to us, and hence the subject matter of the humanities and social sciences' (Pickering 2010: 192).

Arjun Appadurai, whose introduction to *The Social Life of Things* continues to occupy a place of importance in considerations of Material Culture Studies, sketches the concerns of studies of 'things' as follows:

> Even if our own approach to things is conditioned necessarily by the view that things have no meanings apart from human transaction, attributions and motivations endow them with, the anthropological problem is that this formal truth does not illuminate the concrete[!], historical circulation of things. For that we have to follow the things themselves, for their meanings are inscribed in their forms, their uses, their trajectories. It is only through the analysis of these trajectories that we can interpret the human transactions and calculations that enliven things. (Appadurai 1986: 5)

The 'meanings' of things are produced through practices: Whether this results in a 'backgrounding' of the things themselves, or a 'backgrounding' of the practices, suggests a dichotomy of discourse and reality, despite – or as a consequence of – the tendencies Appadurai identifies as 'fetishism' on the one hand, or an 'excess' on the other.

The 'socialization of transactions' Appadurai identifies has, in more recent years, become a 'vitalization'. Here, the 'things' of the title of Appadurai's edited collection are granted (some kind of) agency. Not simply inert vessels for the attributions of meanings, objects and things – and here: materials – must be considered for the ways in which they interrupt, transform, challenge or otherwise disrupt processes of meaning attribution, inserting and asserting themselves. Such approaches[41] are collected under slightly different names, for

---

[41] I use the term 'approach' with care: These are not necessarily (distinct) fields. More importantly, the term 'approach' suggests a manner of getting closer to something (thus more processual than 'field') and as such also evokes Sara Ahmed's 'angles' (cf. Ahmed 2010).

instance, 'Material Ecocriticism' (as in the edited collection by Serenella Iovino and Serpil Oppermann or their 2012 article that preceded the volume), 'New Materialisms' (as in the edited collection by Diana Coole and Samantha Frost) or 'Vital Materialism' (as in Jane Bennett's *Vibrant Matter*, or the online journal of the same name).

In their 2012 article, Serenella Iovino and Serpil Oppermann redress the premises that uneasily rest below Appadurai's 'social life of things' by rejecting the dualistic assumptions that the nature-culture divide rides on. For Coole and Frost, the turn to 'New Materialisms' resists a conceptualization of ontology as abstract, or as absent of 'underlying beliefs about existence' (Coole and Frost 2010b: 5). Their new materialism pushes back on dualist thought, discerning 'emergent, generative powers (or agentic capacities) even within inorganic matter, ... eschew[ing] the distinction between organic and inorganic, or animate and inanimate' (ibid.: 9). At this juncture, they reference the earlier work of Jane Bennett, whose later articulations of what she calls 'vital materialism' have been influential in my articulations of the workings of concrete and plastic.

In *Vibrant Matter: A Political Ecology of Things*, Bennett asks: 'Why advocate the vitality of matter?' Her answer, couched as intuition, reads as follows: 'Because my hunch is that the image of dead or thoroughly instrumentalized matter feeds human hubris and our earth–destroying fantasies of conquest and consumption. It does so by preventing us from detecting (seeing, hearing, smelling, tasting, dealing) a full range of the nonhuman powers circulating around and within human bodies' (Bennett 2010: ix). These capacities of the human, as outlined in detail earlier, do not map evenly across the globe; their effects, whilst also uneven (in particular, through the infrastructures that mitigate the extent of their effects), do not necessarily manifest in the locales, or livelihoods, that gave rise to them.

Jane Bennett's term for conceptualizing her vital materialism is *thing-power*. This term works by mobilizing a remnant sense of wonder for the material world (as in my fascination with the Berlin Wall 'souvenir') alongside a political component of the excesses of meanings (thus) generated:

> *Thing-power* perhaps has the rhetorical advantage of calling to mind a childhood sense of the world as filled with all sorts of animate beings, some human, some not, some organic, some not. It draws attention to an efficacy of objects in excess of the human meanings, designs, or purposes they express or serve. ... The term's disadvantage, however, is that it also tends to overstate the thinginess or fixed stability of materiality, whereas my goal is to theorize a materiality that is as

much force as entity, as much energy as matter, as much intensity as extension. (Bennett 2010: 20, emphasis in original)

With this 'thinginess', or 'fixedness', she recognizes a 'second, related disadvantage of *thing-power*', namely 'its latent individualism, by which I mean the way in which the figure of "thing" lends itself to an atomistic rather than a congregational understanding of agency' (ibid.). My book, with its decided emphasis on materials, cannot circumvent such problems, it must attend to the 'congregational' beyond the 'atomistic'. At the same time, I want to recognize the ways in which language shapes our conceptualization of the world by naming objects or things (the concatenating language of Coole and Frost, trying to approximate new materialisms through lists of terms, comes to mind, see also below). The emphasis on materials in this book is an attempt to bump at the duck-rabbit (or *Kippfigur*) of thing-material: to set a small oscillation of conceptualization, interpretation and mobilization of practices of engaging with that which surrounds us (practices dealing with things) into relations that extend temporally and spatially, with real effects on entities in far-flung times and places.

This emphasis is evinced, for instance, in Serenella Iovino and Serpil Oppermann's introduction to *Material Ecocriticism*:

> Material ecocriticism, in this broad framework, is the study of the way material forms – bodies, things, elements, toxic substances, chemicals, organic and inorganic matter, landscapes, and biological entities – intra-act with each other and with the human dimension, producing configurations of meaning and discourses we can interpret as stories. (Iovino and Oppermann 2014b: 7)

Materials have a great potential for probing relations in ways that push beyond binary modes of thought, such as subject/object or thing/representation. Accordingly, material environments are not 'just' consequences or entanglements of practices of dependence and dependencies (following Hodder 2012) forged through the past, but also extend into an (unknown) future, the yet-to-come. This has an ethical dimension (often coded in terms of 'future generations') as well as a practical dimension ('getting things done', cf. Simone 2004: 22).

Accordingly, I purposefully mis-read the above citation from Iovino and Oppermann in the following way, to stress the *materials* of such an approach:

> Material ecocriticism, in this broad framework, is the study of the way material forms … bodies, things, elements, toxic substances, chemicals, organic and

inorganic matter, landscapes, and biological entities [and how they] intra-act with each other and with the human dimension, producing configurations of meaning and discourses we can interpret as stories. (following Iovino and Oppermann 2014b: 7)

My impetus is not to engage in larger debates of what constitutes 'Material Cultural Studies' or not. I instead recognize that the gap in the field – the *material* of material cultural studies – offers potential to consider (representations of) interactions with the non-human world in a relational manner. Material culture, as 'a system of interrelated objects and social practices' (Sheller 2014: 116), becomes richer, more expansive (spatially and temporally) and more intensive, more inclusive and more … if 'objects' give way to materials.

I use the term 'materiality' to stress the 'materialness' of the materials I examine, that is, the qualities that give rise to consideration of a material *as* material. Materiality has often been thought of in a way that foregrounds mechanical experience: Material objects are then those which take up space, are touchable, weighable, divisible and offer resistance (cf. also Coole and Frost 2010b: 7–8). But materials are not passive, and they break free of their intended uses. If we can talk about oil, for instance, 'breaking loose' or 'sticking to birds' feathers' following an oil catastrophe, then a purely mechanical understanding of the substance oil runs short (cf. Soentgen 2014: 226). The capacity of materials to spread, to detach, disengage, pull free, is a central dimension of our current and ongoing environmental crisis. This is not simply the accumulation of waste, or the bioaccumulation of mercury, issues that undoubtedly warrant attention and will inform the plastic analyses in particular, but also higher levels of carbon dioxide and the thinning of the ozone layer through chlorofluorocarbons (CFCs). Further still, if we can think about oil as fuel derived from fossils – old biological matter – this mechanical view gives way to an expansive imaginative – relational – engagement. It is here where cultural and literary analyses might provide the tools with which we can reckon with the full extent of the material world.

Materiality is defined in Coole and Frost's introduction to *New Materialisms* as 'always something more than "mere" matter: an excess, force, vitality, relationality, or difference that renders matter active, self-creative, productive, unpredictable' (Coole and Frost 2010b: 9). The concatenations of nouns and adverbs are suggestive of the ways in which frameworks and discourses must be exceeded in order to approximate what is meant when introducing new ideas. Jane Bennett, in turn, articulates her use of the term materiality in the following terms: 'Materiality is a rubric that tends to horizontalize the relations between humans, biota, and abiota. It draws human attention sideways' (Bennett 2010: 112). As suggested in the sections 'Modern materials: Manufacturing modernity' and 'Condensing relations: Materialities in the Anthropocene', hierarchical conceptualizations of the world, in which humans occupy a central position, have the tendency to reduce the complexity of the world to its potential to be useful. And, this 'usefulness' is often articulated with respect to very specific human subject positions and very specific human activities.[42]

Reconceptualizing materiality in such a way entails a shift in thinking through the ways we, as humans, relate to our human and non-human co-inhabitants, both biotic and abiotic. Not all processes of materials 'doing their thing' are negative: pickling, brewing, fermenting are all processes where specific configurations are established in order that materials might just go ahead and do whatever it is that they do to, for instance, pickle, brew or ferment. Materials are not only mechanically manipulated, they are also prepared, concocted and cooked. And, following Bennett (ibid.: 115), such materials exert (or *can* exert) potencies on or toward those agents who do the preparations, concoctions and cooking. When we consider agency as extending to non-human co-inhabitants, they are afforded capacities to act and interact beyond our interventions, and without our interaction. Thinking through materiality draws attention to the ways materials forge relations, engender force, exhibit vitality and generally act as agents, shaping our lives as much as we shape them.

My interest in these forms of 'new', 'vibrant' or else-wise modulated materialisms arises from the concern of this book with materials, but also, as the ensuing more analytical chapters will demonstrate, from observations that these materials are *not* passive or otherwise without 'vitality' in the texts I consider. Chapter 2, for example, shifts from depictions of objects floating on, or just below, the pelagic surface to a comprehension of the relations, and toxicities, that the plastic *as material* exudes. In Chapter 3, the properties of the concrete that constitute the (representations of) dams are diffracted with analyses of texts that

---

[42] This way of considering the world is prevalent, even in accounts that grapple with materials. Charles J. Moore, in his account of plastic pollution in the oceans, must realize this in his various calls for action. He tells of presenting the findings of the various teams he works with at different conferences, without much notice being paid to his findings, until Richard Grigg, a Hawai'ian professor, tells him that he needs to show harm. But, nowhere is the subject of this harm made explicit. And yet, as evinced by the path that Moore takes through research done on various sea-based and other critters, and as the narrative build-up of the research he draws on reveals, the harm that needs to be demonstrated is *harm to humans*. So whilst Moore recognizes 'it can't only be about getting baseline measurements to prove and measure its presence'[,] ultimately it's 'about giving credence to the sense that plastic is *doing* something out there, something very possibly unhealthy, something *harmful*' (Moore 2012: 120, emphases in original), *to humans*. I will come back to this point in Chapter 2.

reverberate through various explosions of power and resistance on global and local stages. Chapter 4 suggests that the temporal and spatial dimensions that give rise to plastic's omnipresence as an index of sanitation and hygiene – crucial to modernization – are also the dimensions that give rise to its omnipresence as waste, and as a symbol for the Anthropocene: Plastic's attributes are not shed when anthropocentric utilization ceases. In Chapter 5, the final analytical chapter, the processuality of concrete is stressed: Concrete is not only the material of ruins, but the material through which ruining takes place.

More than a 'cultural history' of concrete and plastic, whacked together haphazardly in the constraints of one book, I use the properties, the functions and dysfunctions, the imaginations and problems of these materials to probe the plasticity of plastic, the concreteness of concrete and, further, the plasticity of concrete and concreteness of plastic – as modern materials of the Anthropocene era.

Towards a method for coming into relations with materialities

How might we use these considerations of materialisms and materiality in interpretative practice? How might we *come to terms* with plastic or concrete (or other materialities) in the *absence* of these very materialities?

Material Cultural Studies, specifically the various newer permutations, offer conceptual frameworks for reconsidering the relations to the non-human world and signpost all manner of philosophical inputs, all the kinds of things which in literary and cultural studies are called 'theory'. The distinction between 'theory' and 'practice' is one that is hardly tenable in horizontalized contexts characterized by 'active, self-creative, productive, unpredictable' (Coole and Frost 2010b: 9) matter. For as Karen Barad points out, 'making knowledge is not simply about making facts but about making worlds' (Barad 2007: 91). This resonates with the notion of invention, again understood not as 'making things up', but as a 'coming into relations'. And invention, in this sense, is intrinsic to cultural artefacts: Some stories, for instance, 'fiction', are 'made up' as worlds of different relations; other stories stress particular aspects or angles (voice, agency, history, etc.) as a way of stressing particular relations. We recognize or learn these things *with* our reading practices, and come to terms with them in very particular ways: Narratology, genre studies, discourse analysis, for instance, and also anthropology, history, literature, political science and so on. These frame our ways of approaching stories, and how we recognize what is happening in them. Other cultural artefacts might depend on context (street, zoo, museum, gallery, rubbish bin, mall) for how we relate to them, and we similarly learn these through specific practices. This book is not concerned with *disciplining* the boundaries of different texts and interpretative practices insomuch as it wants to probe the configurations (and sometimes absences) of materiality.

By shifting away from the emphasis on chains of causality that rely on attributions of agency – particularly to humans – 'intra-action' allows for a consideration of the patternings of agencies across all manner of matter.[43] To put this notion of *plasticity* more *concretely*: plastic and concrete are not dormant materials. They are not made, shaped, and then suddenly stable. They exude their matter, they shift in their shapes and they transform through their functionalities. They extend beyond the physical boundaries, belying any sense of discreteness as object: Their very materiality forges relations and transforms livelihoods.

For a method of materialities, this means closely attending to movement, to shifts, to the way different stuff effects and affects different stuff. Rather than doing this on an abstract level, I bring in Linda Hogan's novel *Solar Storms* to elicit a kind of 'diffractive methodology' that can trace entanglements and dependencies, materialities and intra-actions in a fictional text, attempting to horizontalize the relations between theory and text. A diffractive methodology 'does not concern homologies but attends to specific material entanglements' (Barad 2007: 88). Where reflection is concerned with sameness and mimesis, diffraction concerns itself with differences and relationalities. For concrete and plastic, this means attending to relations without presupposing identity and forgoing attempts to articulate objectivity in terms of authenticity or homology.

In the mode of 'reflection', against which Barad situates their 'diffraction grating' (ibid.: 90), practices of interpretation are often about 'finding accurate representations' and 'about the gaze from afar' (ibid.: 89). Reflection hence pretends to authority as well as authenticity. Both positions – authority and authenticity – rely on ostensibly extra-textual structures to buttress their functionality. Such extra-textual structures are imbued with identity politics,

---

[43] Barad, pushing back on the intuitive assumption of the identity of themselves as author as an individual entity separate from the book they are writing, suggests '"we" have "intra-actively" written each other ("intra-actively" rather than the usual "interactively" since writing is not a unidirectional practice of creation that flows from author to page, but rather the practice of writing is an iterative and mutually constitutive working out, and reworking, of "book" and "author")' (Barad 2007: x). The complexity of this parenthetical aside then gives rise to an array of consequent observations about the dualistic assumptions that gird subjectivity and its location, and about the various entanglements that shaped the production of the book. Infrastructure is not divorceable from structure; context shapes the matters at hand.

and, despite all pretence, find their moorings in the very 'intra-textual' structures that they pertain to be separate from. The cultural critic, scholar or other actor, insofar as they work to establish their authority (or authenticity) as independent of the materials with which they engage, neglect the very complex ways in which their 'objects of study' are the very matter that gives rise to their position in authority, that is, their expertise. Attending to materialities (in the Anthropocene) cannot afford this.

Diffraction also moves away from 'separate entities' such as 'words and things' (ibid.: 89) and towards an 'entangled ontology' of 'material-discursive phenomena' (ibid.: 90). With such an approach, cultural artefacts (like, for instance, literary texts) are not judged in terms of their (presupposed) capacity to reflect on the world – using mimesis to suggest that a particular representation is 'more fitting' or 'more accurate' than another – but instead as co-constitutive of the world. In bringing *Solar Storms* as a test case, I stress the materialities and practices that coalesce around concrete dams.

Because: There is not much *concrete* concrete in *Solar Storms*. The word is entirely missing except for in one description: the floor of a prison cell (Hogan 1997: 291). Otherwise, there are no descriptions of the material, no meditations on the structure of the hydroelectric dams – beaver dams have more attention paid to their material construction than those of the hydroelectric corporation. And yet, a residual sense of the pervasiveness of the *concrete* manifestations of power permeates the text.

Accordingly, the guiding tension of this particular reading is the conspicuous lack of *mimetic representation* of the dams themselves within the novel. An interpretation in the 'reflective' mode will quite quickly reach its limits given this absence, even though the novel is, in my reading, very much engaging with dams, politics and 'representation' (in the political/identity sense, but not to establish a false dichotomy against 'aesthetic representation' either: they are connected). The matter of the dam is absent, at the same time, it is omnipresent. This is important for this book: Even in the more *concrete concrete* texts I analyse in the remainder of this book – or the more *concrete plastic* texts – the materials themselves are absent. There is no actual cement + water + aggregate concrete in the books I allude to, nor in the films, photographs nor many of the other cultural artefacts. No plastic, either (mostly, sometimes the books came in wrappings). Readings *for* materials in the sense that I articulate here are not readings *of* materials in the *concrete* sense of the reflective mode of analysis.

Solar Storms is an activist tale of Native American resistance to power.[44] Not only discursive power, but, indeed, *concrete* power, manifested in concrete dams. The story follows Angel Wings (or Angela Jensen, as she was previously known) as she returns to the places and people of her childhood, most prominently in Adam's Rib. She is becoming reacquainted with the people and ways of her childhood and ancestors when two young men paddle in to the story to announce the construction of dams further to the north. Angel, together with her surrogate grandmother Bush, her biological great grandmother Agnes and her great-great grandmother Dora-Rouge, travel north by canoe in the hope of preventing this construction.

Crucial to my interpretation of *Solar Storms* are the flows and halts of power integral to the story. Investments in dam projects and invested interests in the exploitation of resources extend beyond the locations and boundaries of nation-states to infiltrate various levels of the lives portrayed in the novel. References to judicial frameworks – evoked not only at the close of the novel, for instance, when the peoples of the North are warranted some justice, but also in Bush's evocation of the 'terrorist' later in the novel (Hogan 1997: 288) – shows the uneven patternings of the way violence maps onto state-sanctioned or resistant and activist actions. The long history of colonial power manifests, impacts in the sense of forcible contact, in and through the marks that scar Angel's face, which are paralleled by the marks that scar the face of the earth.[45]

When, as scholars of literary texts, we talk or write of 'things' that recur in texts, we often use very visual terminology: 'imagery', for instance, is often used to render themes. Also, when we talk (or write) of how these 'things' occur in texts, we often use visual codes: foreground, background, highlight and so on. We also tend to do this when we talk of knowledge (enlightenment, obviously, but also show, illustrate, clarify, reveal, etc.). Such language easily gives rise to an interpretation in the 'reflective' mode, concerned with representation as

---

[44] The novel's setting remains vague, both spatially and temporally, although most critics agree that it (cor)responds to the James Bay projects on Cree territory, near the US and Canadian border, following the Quebec government's dam building projects in the early 1970s. The locations and boundaries of nation-states remain vague throughout the text, although the repercussions of colonial treatment of indigenous peoples, manifestations of nation-states, are omnipresent. Whilst the specific cultural heritage of the author – Linda Hogan is Chickasaw, whose lands are in the South-West of the United States – does not map onto the spatial location of the novel's setting, an argument could be made for the systematic oppression faced by all indigenous peoples in settler contexts, for example, the 'Inconvenient Indians' of Thomas King's coinage on the North American continent (see also Chapter 3).

[45] For interpretations of the scarring of the face in the novel, see Stacks (2010) and Sze (2007).

mimesis.⁴⁶ Accordingly, the shifts in scales, space and times narrated throughout the novel are *not* considered here in terms of a function of perspective – although we are only privy to the perspective offered by the narrator, where terminology again stresses paradigms of sight (and epistemological frameworks of Western knowledge). Instead, they are elicited through various objects, elements and flows of materialities that pervade the text. That is, moments of trans-corporeality, 'the material interconnections of human corporeality with the more-than-human world … acknowledging that material agency necessitates more capacious epistemologies' (Alaimo 2010: 2).

Motif, or 'leitmotif' has visual implications, in the sense of a pattern or design, but it is also used in music, for a succession of notes or a brief melodic or rhythmic pattern. Motif, like motive, derives from the late Latin *motivus* from the verb *movere* 'to move'. It is also etymologically related to 'motivation'. I wish to mobilize these related meanings in my reading of *Solar Storms* to nut out specific motifs of the text. By harnessing the 'moving' of the etymology of the term 'motif', I wish to stress three aspects in particular: First, the way in which these motifs shift through scales (of space and time, possibly others), and how mobility is central to such shifts. Second, how 'moving' – together with 'motivation' – stresses the capacity of motifs to effect responses. This is in line with the activist aspects of the novel, and also in the particular ways certain motifs might affect readers. Third, how a motif might not reference a particular word or visual impression evoked by a diadic semiotics of referentiality, but instead as a kind of patterning. This might be a material patterning (e.g. water might be evoked by water, by lake, by tears, by storms, even by fish or canoe) or a patterning of materiality.

I emphasize (*not* 'focus on') three main motifs in *Solar Storms* for the way in which they configure materiality. These motifs probe the flows and shifts of material forms, of entangled vitalities together with the political, ecological and social consequences of uneven development. This is to say, the materialities *not only* engender connectedness and relations, in the form celebrated in material ecocriticism and accompanying discourses, but also, crucially, negotiate the hierarchical consequences of power.

One, water. Obviously, for a novel I am reading in conjunction with dams and hydroelectric power generation, water is a prominent motif in the novel.

---

⁴⁶ If, at times, it sounds unusual that I might avoid some of these verbs of sight in my analysis of Hogan's novel, it is because I am trying to move away from the dominant metaphors of sight as knowledge and towards a language that that attends more readily to materiality (probe, stress, mobilize, emphasize, etc.).

Water is not only symbolic of flow, but also represents mystery and stagnation. It is integral to Angel's understanding of herself *as* water: 'But I was like Agnes had said: Water going back to itself. I was water falling into a lake and these women were that lake' (Hogan 1997: 55). Water covers, shifts, connects and obscures. In the landscape of the novel, it is a prominent element, found in rain, storms, lakes, tears. It is also water that comprises the peculiar agency of 'a circle in the lake where winter ice never froze', the 'Hungry Mouth of Water' (ibid.: 62) into which various people and objects disappear in the text's world. It is the rivers and lakes both fluid and solid, and the clouds of storms (gas). The various manifestations of water, which we as readers are also familiar with from our 'normal lives', are perhaps the most obvious manifestation of shifting materialities in *Solar Storms*.

Two, the wolverine. The wolverine is a trickster figure, never quite human, never quite non-human. The wolverine recurs throughout the novel. Angel notes, in response to her surrogate Grandmother mentioning the wolverine in connection with hunting: 'I didn't ask again what a wolverine was. I'd already begun to think it was an animal with no true description. This time I just listened' (ibid.: 82). Note that rather than offering a conclusive description of the wolverine, the text cautions us to 'just listen': Thus 'underdetermined', the wolverine's capacity to shape-shift is emphasized. The materiality of the wolverine is left open, the figure returns in the closing pages as a constellation in the sky, as well as when a character, LaRue, mistakes his reflection in a window, but catches himself: 'I thought it [his reflection] was too ugly for a wolverine' (ibid.: 350), he defends his mis-recognition. This is, of itself, a beautiful vignette for the deceptive straightforwardness of reflection as a mode of interpretation.

The third motif is a gift, given to Angel by her great great-grandmother: a frog entrapped in amber. Angel takes the frog-in-amber with her on her journey to the north, where it is first stolen from her by her mother (to be recovered after the latter's death) and then succumbs to the flames when the house of the activists is set on fire after being beset by the force of construction workers looking to demolish it. Amber is fascinating: biotic matter trapped in biotic matter, becoming stone (and talisman). Flows of life now solidified, petrified. In the novel, it connects the personal with the political, is stolen and destroyed, and yet returns in the final pages of the book with water and the wolverine. The meaning of motifs extends beyond their material existence within the fictional world of the novel, brought together at its close.

The materialities of the three motifs – water flowing, freezing and swelling behind dams, river and lake, tears and rain; a trickster figure, never quite human, never quite non-human, a shape-shifter; the frog-in-amber, biotic matter-turned-stone – are themselves motifs of connections and shifts, and of moments 'frozen' and flowing, forging a material sensibility. Together, the motifs evoke a *plastic* sense of relations that emerge through *concrete* configurations, configurations that are themselves open to change.

* * *

In the following chapters, I think with and through materiality to be attentive to relations and effects – and to the differences that matter – rather than (just) the things themselves. It is through this shift in emphasis that the materials come into effect: materials that themselves forge relations and produce effects. Thinking about materials *sets* in here, the *concrete* idea that frames this book in its *plastic* forms and shifts.

2

Plastic Pacific

Introduction to the Plastic Pacific

On the shores of a Hawai'ian beach, Kamilo Beach on Big Island, was where plastiglomerate was 'discovered' for the first time. 'Discovered' in scare quotes, because the material relations that give rise to this strange material of the Anthropocene were there before scientists and artists stumbled across it, photographed it, removed it for study.[1]

But, before we get too attached to this story, let's try telling it a different way: Plastiglomerate is an *invention* of the Pacific (invention, from *invenire*, see also 'Introduction'). It is in the Pacific that humans have come into specific relations with plastic, and it is different attempts to imagine and trace these relations of plastic emerging through and within this space – an expansive space, one third of the earth's surface – that I wish to examine in this chapter. A further caveat: In this chapter, plastic is mostly waste. It is excess, externalization, toxicity and, for some, obsession; sometimes it figures as a mass material, other times it is imagined as discrete objects.

One prominent, perhaps the most prominent, manifestation of plastic in the Pacific is not the plastiglomerate evoked above, but rather as mass comprising the 'Great Pacific Garbage Patch' (GPGP).[2] This is particularly evident in a selection of book-length investigations into plastic relations and pelagic waste published by journalists. These books are perhaps best classified as 'science

---

[1] In a discussion about plastic things with Baldeep Kaur, they pointed out how plastiglomerate faces the fate of many a colonial artefact, to be removed to the Global North for knowledges' sake. The inclusion of Joshua Franzos's image as Figure 2.1 – which shows a plastiglomerate on display in a museum – tries to approximate the complicated entanglements of material knowledges.
[2] I wonder about the 'garbage' in the moniker: the US/Northern American framing of the issue – which becomes evident in this chapter – is there right from the beginning with this word. Other Englishes, as Alice Te Punga Somerville has also noted, would use terms like 'rubbish', 'litter' (cf. Te Punga Somerville 2017: 321) or perhaps even 'waste', 'debris'. But, because it is 'garbage', we can have the cute abbreviation GPGP.

Figure 2.1 'Plastiglomerate' by Joshua Franzos. From the Exhibition 'We Are Nature' (Oliveira et al. 2020: 34). CC BY © Joshua Franzos.

communication', though they also have overlaps with travel writing, nature writing, investigative journalism, biography and other generic conventions from within non-fiction. Across these books, a concern with representation becomes evident, and a predilection for two specific images emerges: the rubber ducky[3] and the island. Both these images are examined in what follows.

My interest in the rubber ducky does not seek to pass judgement on its adequacy as an image. I rather observe that it is an image that recurs across a range of texts, though particularly in the book-length science communication investigations. With the rubber ducky, discussions of the repercussions of toxicity, together with considerations of children and of birds, coalesce in a 'familiar' object. My interest in materiality, as outlined in the introduction, stretches beyond the various objects made of specific materials, and it is the ways in which these discourses stretch *beyond* and condense *through* the evocation of specific ideas that I focus on in what follows.

---

[3] My earlier thinking through these ideas, which I can trace to a talk I gave in early 2016 as part of a conference held at the University of Vienna called *Maritime Mobilities*, has since been published (Crane 2023a), and I would like to thank the audience as well as our host, Alexandra Ganser, for input on the talk and feedback on its written form. In that piece, I think more broadly about birds and specifically about rubber ducks *and* albatrosses. I have since learnt from reading Max Liboiron to not reproduce these images, that they do not sit with more-than-human ethics. This chapter thus does slightly different work from that version.

Birds can forge connections between disparate sites. Most birds fly; many birds are at as home on water as on land. Their history as an index for anthropogenic environmental destruction forges connections to Rachel Carson's *Silent Spring* (1962). They also stand for anthropogenic species extinction. The class of Aves, birds, offer three well-known species that have become extinct in their relations with human beings: the dodo (*Raphus cucullatas*) of Mauritius, the great auk (*Pinguinus impennis*) of the north Atlantic and the passenger pigeon (*Ectopistes migratorius*), once one of the most common bird species on the planet. But there are many other bird species, including some ducks, which have also gone extinct: the Mariana mallard (*Anas oustaleti*), Finsch's duck (*Chenonetta finschi*) and the Auckland merganser (*Mergus australis*).[4]

Ducks, in what follows, are however mostly plastic, and known as rubber duckies. The rubber ducky works in (at least) two ways in the 'Plastic Pacific'. It has, firstly, reached iconic status as an index of pelagic waste. Rubber duckies, as flotsam, draw attention to the shifts and changes of the currents of the oceans, forging connections between otherwise disparate sites, revealing even the swirls and entanglements of the gyres. They bob through children's books, like Eric Carle's *10 Little Rubber Ducks*; they grace the cover of Curtis Ebbesmeyer and Eric Scigliano's *Flotsametrics and the Floating World* and figure in the titles of other books (*Moby Duck, Slow Death by Rubber Duck*).

Secondly, the rubber duck emerges as an icon, if not index, of plastic toxicity (as its use on posters and banners in protest of the polyvinyl chloride spills of East Palestine (Ohio, USA) in 2023 suggests). Stacy Alaimo has argued that the 'persistent (and convenient) conception of the ocean as so vast and powerful that anything dumped into it will be dispersed into oblivion makes it particularly difficult to capture, map, and publicize the flow of toxins across terrestrial, oceanic, and human habitats' (Alaimo 2012: 447).[5] I argue that attempts to redress such representational issues are being made with the rubber ducky in the texts I examine, and innocuity and innocence play an important role in how this image works. Indeed, the disjunct between toxicity and bath-time toy is elaborated most explicitly with respect to Rick Smith and Bruce Lourie's *Slow*

---

[4] I foreground these examples to forge a tentative link to rubber duckies. All three duck species originate from the Pacific (though note also the taxonomic status of the former two as clearly distinct species is open to debate).

[5] I have used the term 'toxins' where the authors use the term: mostly, 'toxins' maps onto both toxins and toxicants, a distinction Max Liboiron, Manuel Tironi and Nerea Calvillo make to stress that differences in providence (toxins exist in plants, animal cells and minerals; toxicants are manufactured, or at least introduced by industrial interventions in different places) as well as scale (the former are present in miniscule amounts, the latter at much larger scales) (cf. Liboiron, Tironi and Calvillo 2018: 333–4).

Death by Rubber Duck, where the form of the duck gives way to the material of its plastic as it begins to seep through the boundaries of its form into biotic matter.

The other idea that shapes this chapter – islands – draws on archipelagic thought to stress the non-dualistic, non-hierarchical, non-linear engendered through and with 'chains of islands'. Shu-mei Shih argues that archipelagic thought brings 'submerged or displaced relationalities into view' (Shih 2008: 1350) and pushes back against 'the meaning of comparison as the arbitrary juxtaposition of two terms in difference and similarity, replacing it with comparison as the recognition and activation of relations that entail two or more terms' (ibid.). Thinking 'Plastic Pacific' with the archipelago, I follow in particular the work of Alice Te Punga Somerville and Craig Santos Perez. I turn to Perez's poem 'Age of Plastic', which brings the concerns elicited in conjunction with various rubber duckies together with archipelagic sensibilities, at the close of the chapter. I trace this poem's embeddedness in a particular volume of poetry as well as in Perez's academic writings, in order to work out the archipelagic qualities of the 'Plastic Pacific'.

'Debris in the Pacific Ocean', Alice Te Punga Somerville contends, 'both marks and stands in for a series of historical moments in which an original item was produced, discarded, became debris, and travelled' (Te Punga Somerville 2017: 327). She argues that the 'plastic bottle floating on the crest of the wave' is not solely about this moment, that is, the acts of looking and floating, but also 'about the circumstances that led to it' (ibid.: 328). This is a recognition of the stretchiness of the material relations across time and place, which then condense in the plastic bottle. Plastic's capacity to preserve itself – which I examine in more detail in Chapter 4 – is one way plastic *does* plastic. Similarly, findings that plasticizers are not molecularly stable, that is that 'they tend to leach and off-gas' (Davis 2022: 87), is, well, plasticky things doing exactly the thing they are supposed to do: make things more plastic. Such characteristics are traced in this chapter as toxic relations.

The flows of plastic, here in the Pacific, require more complex readings than can be offered by overtly dualistic models. An attempt to read the relations of modernity to waste as entirely 'what is left behind' or 'what is excluded' is resisted here and in the other chapters, for such a reading in terms of externalization suggests temporal and spatial segregation. For the moment, I sit with the ways in which plastic's promise of modernity has, for some time now, given way to a pledge of sludge, an inundation of waste. The promises and inundations are uneven: Some places and some people have to deal with the dubious aftermath of copious consumption to greater extents than others.[6] Heather Davis develops the idea of transmission to reckon with this idea. The term 'transmission' shifts away from the potentially beneficial consequences that the close cognate 'inherit' elicits, instead patterning across the globe and time unevenly and, crucially, unfairly: 'Differentiated from inheritance, where plastic becomes the problem of those who invented and benefit from it', Davis argues, 'transmission describes the imposition of plastic: its legacies on multiple peoples, largely racialized and poor, who deal with the intergenerational effects of plastic but are not responsible for its emergence or proliferation' (Davis 2022: 5–6).

In this chapter I probe plastic as waste in conjunction with the particular site of the Pacific Ocean. It is an impressive site of relations, comprising almost a half of the world's water surface and one third of the globe's entire surface, and connecting to all other oceans of the world. It is also an impressive site of environmental concern, including nuclear testing, bombs and fallout.[7] Many of the island countries that comprise the Small Island Developing States (SIDS) and the Alliance of Small Island States, coalitions formed to address issues of sustainability and higher sea-levels on the international level (e.g. lobbying the United Nations), are situated in the Pacific. Rapa Nui (Easter Island), the Galapagos Islands, the Midway Islands, the Marshall Islands – sites that have become iconic for specific environmental and scientific issues – are all situated in the Pacific. Indeed, this chapter's title 'Plastic Pacific' shows homage to Elizabeth DeLoughrey's work on the 'Nuclear Pacific' (cf. DeLoughrey 2012). And, further, the Pacific comprises many island peoples whose histories have rendered them open (and exposed) to sea-faring visitors, both human and not (cf. e.g. Matsuda 2016). The Pacific, far from being an 'away' to which waste can be somehow thrown, is shown to bring all forms of plastic to the fore: metaphorically, indexically, literally ... as rubber ducks, as islands, as becoming-plastiglomerate.

---

[6] The debris of modernity is *constitutional* in two further Pacific imaginings of rubbished island spaces: In Waanyi writer Alexis Wright's novel *Carpentaria*, plastic accrues with other discards to form the island off the coast of north-east continental Australia upon which character Will finds himself (cf. Wright 2007). Will's environmental interventions – monkey-wrenching an open cut mine – do not spare him post-cyclonic abandonment on the 'left overs' of society. Taiwanese writer Wu Ming-Yi has one of the characters from his novel *The Man with the Compound Eyes*, Atile'i, similarly stranded upon a Pacific trash heap. After been jettisoned from his Pacific society, Atile'i's seafaring skills means he survives, if only barely, an ordeal on the open seas, to be washed up on an island of debris (cf. Wu 2014). I address the archipelagic qualities of these texts in more detail elsewhere (cf. Crane 2023c).

[7] Many nuclear powers have conducted tests in the Pacific (the United States, the UK, France); the bombs on Hiroshima and Nagasaki; the 2011 earthquake and tsunami events that have left a wake of slow fallout (cf. also DeLoughrey 2012).

Travelling troubles, mobile materials: Object(ion)s to study

In his book *Visit Sunny Chernobyl: Adventures in the World's Most Polluted Places*, Andrew Blackwell affords the accumulation of plastic in the Pacific continental status in the chapter title 'The Eighth Continent: Sailing the Great Pacific Garbage Patch'.[8] The GPGP thus becomes a sort of anti-tourist destination (as well as becoming 'grounded' in the idea of continents – I will return to this later). But, of all the sites Blackwell tours, this one is the trickiest to actually visit: As Blackwell notes, the mass of plastic in the Pacific is not an island. He makes this point with some insistence, repeating the negation twice: 'Let's nip this in the bud: It is not an island. / I'd like to say that again. It's not. An island' (Blackwell 2013: 118). In doing so, he forges a connection through the emphatic negation.[9] The term 'island' he unwillingly conjures suggests a containable space, a discrete object, a stable entity (at least on human timescales, and for most of the time). As Susan Freinkel suggests, 'A floating trash island would be a far easier problem to take care of' (Freinkel 2011: 130).

Even if the 'Great Pacific Garbage Patch' is not a problem that lends itself to visual representation, it nonetheless remains a problem of representation. Blackwell accordingly notes that the GPGP 'isn't a visual problem, and this conflict between the reality of the problem and its nonvisual nature is at the root of the plastic island misconception. A metaphor is needed, a compelling image to suggest the scale and the mass of the problem' (ibid.). This is evinced in his attempts to come up with a metaphor for the phenomenon (an ecosystem (cf. Blackwell 2013: 119) and, later, a galaxy (ibid.: 148)).

Keep in mind that images do not depend on mimetic effect.[10] Migrating through time and space, images and icons produce and reproduce, garner and gather, responses. With respect to the Plastic Pacific, I note an abundance

---

[8] Of the task he set himself, Blackwell observes:

> My mission was to find *the world's most polluted places* ... . But instead of finding degraded ecosystems that I could treat *as though* they were beautiful, I was just finding beauty. The Earth had gotten there first. I went looking for a radioactive wasteland and found a radioactive garden. I went looking for the Pacific Garbage Patch and found the Pacific Ocean. (Blackwell 2013: 154)

More context reveals these things to be a bit more complicated: note, for instance, Kate Brown's work on de-bunking assumptions that the radioactive wastelands around Chernobyl are particularly garden-like (Brown 2019).

[9] This is ironic process theory in action: The popular example is the demand 'do not think of a pink elephant', where the idea is that the first thing you think of is, of course, a pink elephant. And, by evoking islands twice, we are already on our way to archipelagic thinking.

[10] No human was contemporaneous with dinosaurs, and yet collectively we have a rich inventory of images and imaginations of dinosaurs to draw upon (cf. Brosch and Crane 2014). See also the discussion in the introduction to this book.

of images of birds – in particular, rubber duckies. The flows of plastic in the birds of this chapter offer a corrective to the imagination of the isolation of the Pacific, rendering disparate spaces and practices connected, working to counter dominant notions of the islands of the Pacific (and elsewhere) as distinct, static, far-flung ('away') entities.

The images, particularly of the birds, are relational. Plastic as waste in the oceans constitutes a flow of materials, connecting seemingly far-flung sites. It also constitutes a flow of risk, of toxicity, that often manifests in the images, visual and written, of birds in some of the accounts I analyse. The birds themselves forge connections, shifting between different spheres (land, sea, sky). Although such images might appear discrete, the forcefulness emerges through the multitudes of connections in the texts that follow, citing and expounding on each other.

Accordingly, Mark Jackson argues that 'plastic presents itself as an ethical concern for us precisely because it cannot be interiorized within the living horizons of the planetary bird, or similar earthly life processes' (Jackson 2012: 216). This response to the continuing presence of objects made of the material of plastic – Jackson is referring to the Laysan albatross images here – is to foreground his conclusion in terms of ethics and processual grounds: 'With the junk-space proliferation of zombie matter like non-biodegradable plastics (one could add to plastics the deleterious effects of enriched nuclear materials, for instance, or persistent organic pollutants like DDT, etc.), and the inability to adjudicate its limits, we neglect the temporal grounds of our now gyrating and coagulating, non-interiorizing waste' (ibid.: 220). The ethical impetus Jackson proposes, through which putrefaction and decay act as cornerstones for shifts in materiality and relations, has been influential in thinking through this chapter.

However, Jackson neglects to consider two crucial aspects: firstly, in evoking the 'carboniferous efflorescence of stored, solar matter, released in imaginative material possibilities for and of thought, is tens of millions, no, hundreds of millions of years old' (ibid.), Jackson neglects to consider that the plastic he terms 'non-biodegradable' (ibid.) might similarly require a temporal framework that far exceeds human life-scales. Heather Davis suggests that this is due to plastics lacking relations with the world around them – 'to decompose, plastic needs to have a formal relation to the world around it' (Davis 2022: 38) – or more insistently: 'Synthetics actively deny the relations in which they are bedded' (ibid.: 46). I want to think through a different aspect, namely, that in considering the full scale of plastics, we might attempt to grapple with the repercussions of the fossil in the future. That is, the timelines of degradation of plastics stretching into the future might mirror those of the 'millions of years' (to quote Jackson) that have gone into their creation.

The problem, rather than (non)biodegradability, is what plastics do to us and our co-inhabitants in the meantime. And this comes to the second point: In suggesting plastic cannot be interiorized, Jackson does not account for the capacity of plastic to adsorb toxicants and to exude them. Plastic, in an account such as his, remains an object, material manifest in form (rather than as fluid, if, for the purpose of argument, we pretend that the two are somehow oppositional). If the visual representations of plastic ingestion suggest that the objects remain intact (and 'not interiorized'), this is as much a function of the limits of the form of representation than the processes of consumption that are thus depicted. Plastic's timelines are different to our own; plastic's capacity to exert its plastic-ness is not necessarily reliant on its discrete 'identity' as objects. Plastic, in other words, needs material thinking.

Concatenations of plastic

Susan Freinkel's book *Plastic: A Toxic Love Story* subtly shifts from a concern for the accumulation of plastic as debris, fragmented but resisting biodegradation, to a more trans-corporeal comprehension of the toxicity of plastic as material. Her book as a whole is not specifically concerned with the Pacific, although the Plastic Pacific figures largely in her fifth chapter, 'Matter Out of Place'. In this chapter, a cigarette lighter functions as a *Denkobjekt* (roughly: thinking object, see 'Introduction'), forging connections across time and space, as the focus shifts towards the toxic relations of plastic. Freinkel's cigarette lighter is an image from an unnamed researcher's trawled plastic haul from 'a remote stretch of the Pacific Ocean northeast of Hawaii' (Freinkel 2011: 115). Most of the matter they recover 'was unidentifiable bits and pieces' (ibid.); however, this lighter remained (remarkably?) intact, including an address and phone number, which Freinkel (based on the western seaboard of the United States) traces across to the other side of the Pacific in Hong Kong. The lighter has traversed, physically, then metaphorically, a remarkable spatial distance from Hong Kong to a stretch of the Pacific Ocean northeast of Hawai'i, and then, as image, even further to the writer (and, finally, to her reader). It has also, as Freinkel discovers, traversed at least seven years, mostly unscathed by the passage of time.

Using a somewhat contrived transition resting on the subjunctive 'could', Freinkel then forges a connection between the cigarette lighter and the images of Laysan albatrosses, which were for a time omnipresent: 'One place that a castaway lighter *could* land is on Midway atoll ... also home to the Laysan albatross. ... Some 1.2 million of the birds nest on the atoll, and almost every one has some quantity of plastic in its belly' (ibid.: 116, emphasis added).[11] This forged connection is also an entanglement a dependence and a dependency (cf. Hodder 2012). Freinkel suggests:

> The Laysan albatross's proximity to the garbage patch has made it the poster child for the hazards of plastic marine debris. But the birds are hardly the only animals affected by the increasing presence of plastics in the deep ocean. Other sea birds, fish, seals, whales, sea turtles, penguins, manatees, sea otters, and crustaceans have reportedly ingested or become entangled in plastic debris. (Freinkel 2011: 132)

Entanglement, ingestion; debris, material. The tension between object and material is evident throughout Freinkel's book.

A trope favoured by writers on plastic is to list all the various objects made of plastic that their reader might encounter, or that the writer themselves has encountered: a concatenation of plastic, if you like. The list, thinking with the material further, evokes the long chains of polymers that comprise plastics. Freinkel claims that 'we all live in Plasticville' (ibid.: 1), and, after abandoning an attempt to go an entire day without touching anything plastic, she decides to write down everything made of plastic that she touches instead. In addition to the pen and notebook she uses to make the list, her first forty-five minutes include a further forty-three items as various as 'mattress', 'cellophane wrapping of tea', 'fleece sweatshirt', 'eyeglasses' and 'tub containing cat food' (cf. ibid.: 2). Her list from that day comprised 196 distinct items, she reports. The edition of her book that I consulted for this chapter includes a similar list in a barely visible glossy typeset, with items listed in the singular and plural, on the dust jacket.

A list of objects she encounters on Kehoe Beach, near San Francisco on the US Pacific coast, comprises another such list. In reflecting on her beachcombing

---

[11] Although it is unclear exactly which photographs Freinkel responds to – she mentions neither Jordan, nor Liittschwager/Middleton – her response is coded in terms of a juxtaposition of different kinds of matter: 'Every carcass seems a mockery of the natural order: a crumbling bird-shaped basket of bleached bones and feathers filled with a mound of gaily colored lighters and straws and bottle caps. The birds are dissolving back into the ground; the plastics promise to endure for centuries' (Freinkel 2011: 118). The ingestion of plastic by the Laysan albatrosses leads Freinkel to query human digestion of plastics in an anthropocentric move common to many accounts of plastic (and a chief concern of Charles Moore and Cassandra Phillip's accounts of attempts to direct media and political attention to the accumulation of plastic in the Pacific in *Plastic Oceans*, explored in detail later).

practices, Freinkel notes that it takes her a while to 'recalibrate [her] inner treasure hunter' (ibid.: 127), to focus not on the objects previously revered, such as pretty stones and shells, but on 'all the junk' (ibid.). The assemblage of objects washed up on the shores, she later claims, has a global reach: 'Whether they are working a beach in Chile, France, or China, volunteers [for the Ocean Conservancy's international beach-cleanup] inevitably come across much the same stuff' (ibid.: 128). Some objects garner more attention: The awe in the presence of an 'olive green tag that was traced to a U.S. Navy bomber shot down more than ninety-six hundred kilometers away – in 1944!' (ibid.: 117) is the sublime of distance, spatial (9,600 kilometres, over a quarter of the circumference of the earth) and temporal (spanning some sixty years, more than two human generations).

According to Freinkel, the lighter was one of five hundred pieces of plastic found in one chick's stomach: The sublime of distances overcome is conjoined with a sublime of mass. In her account, it is the message in the bottle, an ages-long practice, that connects 'the vast world' and distances 'suddenly seem much smaller and more negotiable' (ibid.: 115). But spaces are also connected through other practices. She accordingly mobilizes her affective response to photos of the Laysan albatrosses with ingested plastic to forge links between the accumulation of plastic (in the form of debris) and the accretion of toxicants (bio-accumulating in the bodies of animals, human and non-human). Ending her chapter with an ekphrastic description of a *Life* cover depicting 'convenient disposables' appearing to be 'magically suspended midair' (ibid.: 139, the reference is to the 1955 cover), Freinkel imagines what is not in the frame, that is, the pile of objects accumulating after disposal. These are, her analysis suggests, the objects and fragments that accumulated in the Plastic Pacific, and in the guts of the Laysan albatrosses. But plastic also accumulates in the bodies of animals, human and non-human, not in recognizable forms, but as hormone-disrupting chemicals, through the chains of ingestion (cf. e.g. ibid.: 137). Freinkel's explication of the endocrine disrupting PCBs (polychlorinated biphenyls) is brief, but points in the direction of a toxicity that qualifies our relationship to plastic beyond that of objects. Plastic no longer bobs (on, or just below) the surface of the oceans, it reaches into the oceans, exercising a malignant agency.

This is, in Jane Bennett's terms, a vital materialist attitude towards plastic. In *Vibrant Matter*, Bennett explains that the vital materialist can respond to such concerns as follows:

First, by acknowledging that the framework of subject versus object has indeed at times worked to prevent or ameliorate human suffering and to promote human happiness or well-being. Second, by noting that its successes come at the price of an instrumentalization of nonhuman nature that can itself be unethical and can itself undermine long-term human interests. Third, by pointing out that the Kantian imperative to treat humanity always as an end-in-itself and never merely as a means does not have a stellar record of success in preventing human suffering or promoting human well-being: it is important to raise the question of its actual, historical efficacy in order to open up space for forms of ethical practice that do not rely upon the image of an intrinsically *hierarchical* order of things. (Bennett 2010: 12, emphasis in original)

This *hierarchy* of things enacts specific constraints on activists working to explicate the issues of the Plastic Pacific: the need to demonstrate harm, to humans, as Charles Moore and his team discovered, and as described in *Plastic Oceans* (see below).

The concatenation of objects finds a horrible objective correlative in the widely circulated photographs of Layson albatrosses, mentioned by Freinkel (and others). Chris Jordan's series 'Midway: Message from the Gyre', a series of photographs of dead birds taken on the Midway Atoll taken from 2009 onwards, reveal the stomach contents in the dead birds: a mosaic of colour; a collection of plastic. Subdued remnants of bird share the frames with lurid pieces of plastic emitting from their guts. The juxtaposition is startling. Feathers, bones, beaks, upon a bed of sand, all 'earthy' tones, shades of grey and brown, in contrast with the (lurid) greens, blues, reds, oranges and yellows of manufactured plastic. Sometimes, these pieces of plastic are readily identifiable as objects; often enough, they are only identifiable in terms of their material, that is, as plastic. Each photo captures individual devastation; together, they speak to a larger, systematic, concern.

These features, particularly the scientific gaze evoked by the all-encompassing perspective of the photograph taken directly above, suggest a style that stresses documentation over other considerations. Documentary photography entails attitude or stance towards the subject, more than any issues of aesthetics, seeking to depict subjects in an objective or candid manner, operating in the mode of the chronicle or archive. Photographic documentation, in turn, is associated with forensic procedure (for instance, in providing evidence for physical harm inflicted on bodies). Documentation in photography has a long tradition, almost as long as photography itself, reaching back to the works of Jacob Riis, Lewis Heine and Dorothy Lange, for instance. All three photographers sought to elicit social change through the depiction of specific subjects. Mark Jackson notes that 'Jordan's photographs … raise such questions [of material and cultural processes] through visualizing the intimacies of distant death precisely in the apparent juxtaposition of matter' (Jackson 2012: 209). Documentary photography might, then, not be seen as a 'it-was-so' (to paraphrase Barthes' *ça-a-été*) but as a 'it-could-be-otherwise', where the deictic reference of the photograph reaches not (only) into the past but rather extends into the present and the future.

The photographic works must be considered, then, not only for what they depict, but for how the specific frameworks of the photographs elicit response, that is, how contexts work to frame the photographs. In interview with Bill Hickman, Jordan emphasizes photography's capacity to bear witness (cf. Hickman 2012). Gillian Whitlock, in her work on testimonial transactions in postcolonial life narratives, draws attention to the metaphor of bearing witness: the dialogic and rhetoric of testimony pulls the addressee into the account, transferring a weight of responsibility and affect (cf. Whitlock 2015: 8). Here, bearing witness is complicated by the non-visual nature of the subject matter. In Jordan's case, the photographs are accompanied by a brief 'About' text on the website. Jordan uses the imagery of mirror and reflection in 'About This Project' to suggest that the birds are 'an appallingly emblematic result of the collective trance of our consumerism and runaway industrial growth' (Jordan 2011). 'Like the albatross', the text continues, 'we first-world humans find ourselves lacking the ability to discern anymore what is nourishing from what is toxic to our lives and our spirits. Choked to death on our waste, the mythical albatross calls upon us to recognize that our greatest challenge lies not out there, but *in here*' (ibid., emphasis in original). The images of the albatross, then, forge relations across distances and across the interior/exterior boundary.

Chris Jordan's photographic project has a famous precursor: David Liittschwager and Susan Middleton's 'Shed Bird' image (which is often displayed as a singular image comprising two photographs). In their photos, the birds are laid against a white background – clinical white, evoking hospitals and by extension the morgue – the plastic, in pieces, testifies to a pathology. This pathology is one of consumption, which I turn to now, and one of toxicity, as I will discuss in more detail later.

The pathology of consumption leads, it appears, to the death of the albatross. But not quite: Consumption does not 'lead to' the death of the albatross, the image does not allow for the distancing a path or trajectory this phrasing ('lead to') entails. The pathology of consumption *is* the death of the albatross.

The accompanying text contextualizes the image as follows:

David and I were shocked to see massive amounts of marine debris strewn across the beach: floats, plastic lighters, plastic shoes, toys, thousands of liquor bottles, television tubes, laundry baskets, light bulbs, plastic containers of all sizes and shapes, medical waste – the flotsam and jetsam of civilization. Ropes, nets, and lines present entanglement hazards to wildlife, including seals, turtles and birds .... I know we were in one of the most remote places on earth, but suddenly civilization seemed pressingly close. (Liittschwager and Middleton 2005: 129–31)

This list of objects, another concatenation of plastic, suggests the broad reach of plastic's entanglement in modernity. Whilst the list does elicit a sense for the multitudes of objects made of plastic that pervade our lives, in this list they remain objects: clearly identifiable in the sense of (past) function and form.

The caption to Liittschwager and Middleton's image reads 'the mosaic of *death*'. This 'death' not only references the death of the young albatross; in my reading, the lack of qualifier or pronoun suggests not just this individual death but death on a broader scale. With the indefinite death of the caption, the image of the contents of the young albatross' stomach proffers a metonymical reading, the death *both* of a specific organism (whose representational impact is key to the affectual dimension of the image) *and* of biotic life in general (humans and otherwise).

The albatrosses, amongst other birds of the Plastic Pacific, thus function as a symbol of the global – of disparate sites and their connections. Birds of flight forge imaginative and real connections between spaces in ways terrestrial animals might not. Marine birds, like the Laysan albatrosses and (not-so-rubber) ducks connect the terrestrial with the aquatic, and they also forge connections between different terrestrial areas. And it is the terrestrial that dominates human understandings of the globe.[12] As more than one article on the Laysan albatross images suggests, these birds are for the oceanic world what canaries are for

---

[12] The globe is dominated by a territorialized understanding of space. Humans are first and foremost land-dwelling species; our abodes are usually constructed upon *terra firma* (and even if they are not, they still use materials from the earth). Maps foreground the static: the earth, its structures and static biota (e.g. forests, i.e. biota that are stationary *as mass*), but these are all only static at the human scale. Even continents shift. These are the forms we are trained by the images to grapple with, ahem, the *earth*. As Charne Lavery points out, the floors of the ocean are by no means conclusively mapped, mostly because satellites cannot penetrate the surface of the water (Lavery 2015: 26). This is an effect of the visual. Satellite imagery cannot penetrate water, so whilst isolated regions of (the) earth are incorporated in visual representations – becoming part of the body of maps and the body of knowledge they represent and reproduce – the bottom of the floors of our oceans retreat, rescind. The 'depths' of our comprehensions of (our entanglements in) the world are, in some spheres, rather shallow.

the subterranean world.[13] The trans-corporeality of plastic, the 'material inter- and intra-connections between living creatures and the substances and forces of the world' (Alaimo 2016: 113), emerges most forcefully with the increasing recognition and, crucially, articulation of the toxicity of the material through images, such as the rubber duck.

Flotsam and narratives of risk (rubber duck i)

I turn to Donovan Hohn's *Moby-Duck: The True Story of 28,800 Bath Toys Lost at Sea* in order to probe a particular phrase from the quote earlier: plastic waste as 'the flotsam and jetsam of civilization' (Liittschwager and Middleton 2005: 129). Hohn's account is best characterized as nature writing meets travel writing.[14] The book is concerned primarily with the rubber duck, or, more specifically, the yellow ducks, red beavers, green frogs and blue turtles that comprised a load of flotsam that has been traversing the Pacific since a spill in January 1992.[15] In the prologue, Hohn accordingly notes: 'I'd never heard of the Great North Pacific Garbage Patch. ... I just wanted to learn what had really happened, where the toys had drifted and why. I loved the part about containers falling off a ship, the part about the oceanographers tracking the castaways with the help of far-flung beachcombers' (Hohn 2012: 3). He is not alone in his captivation: amongst other stories, he also references Eric Carle's *10 Little Rubber Ducks*, an infant board book in Carle's characteristic aesthetics familiar from *The Hungry Little Caterpillar*. As in Carle's book, Hohn renders the rubber duck iconic, foregrounding it in the title and throughout the book.

Hohn takes his readers on trips to visit Curtis Ebbesmeyer (one of the 'oceanographers tracking the castaways', often accredited with coining the phrase 'Great Pacific Garbage Patch', cf. also Ebbesmeyer and Scigliano 2009).

---

[13] I refer here to a plethora of weblogs, too disparate and lengthy a list to collate. Teflon toxicosis, when pet birds' lungs haemorrhage and cause them to drown after exposure to Teflon (e.g. from coated pans), is a modern-day version of the fate of the miner's companion species: Rick Smith and Bruce Lourie elaborate on this phenomenon in detail (cf. Smith and Lourie 2009: 85–8).

[14] This generic description could also be applied to Blackwell's *Visit Sunny Chernobyl*, cited earlier. On the overlap between the two genres, see also Crane (2019a). 'New' nature writing, as some have called such writings, brings its own debates (cf. Cocker 2015 and Macfarlane 2015). The didactic qualities of many of these texts make me also think of them in terms of science communication, as mentioned earlier, but let's just note their hybrid qualities for the moment and return to the analysis.

[15] In his collection of poetry *Plastiglomerate* (2020), Tim Cresswell traces the movement of some of the ducks, beavers, frogs and turtles, pairing (documented) sightings of these with *Tripadvisor* comments ('Friendly Floatees/*Tripadvisor*'). These rubber duckies have a pervasive presence.

We also accompany him on several boat trips, with the book structured to recollect Herman Melville's *Moby-Dick*, as the title already suggests. In Hohn's account, the Pacific is thus rendered interconnected with other sites, and oceans, by means of the various trajectories he follows (which he refers to as chases). In the 'Fourth Chase', which takes Hohn to the sites of manufacture of the rubber duck in China, Hohn dwells on the cultural significance of the rubber duck, and also its iconic yellow. Hohn suggests that the 'commercialization of childhood and the infantilization of animals' coincided with some other auspicious events to give rise to the rubber duck:

> In 1871, a printer from Albany, New York, named John Wesley Hyatt added nitric acid to pulped cotton, thereby inventing celluloid. In 1873, the first Pekin ducks were imported to the United States from China. And in the 1880s, bathtubs began appearing in middle-class homes along with indoor plumbing. Celluloid eventually evolved into the plastics industry. The Pekin duck eventually became the preferred species of American duck breeders, making yellow ducklings a familiar symbol of birth and spring. (Hohn 2012: 224)

The rubber duck, Hohn's brief narrative goes on to suggest, was inserted into bath-time in order to distract small children from playing with their genitals (in a story that entangles physical health and cleanliness with perceived mental hygiene risks). Quite an in(ter)vention.

Drawing on the often-quoted passage on 'Plastic Man' by Victor E. Yarsley and Edward G. Couzens, in which the 'Plastic Man' is kept safe, kept clean, kept pleased (even kept dead) by various plastic objects, Hohn argues: 'Here, then, is one of the meanings of the duck. It represents this vision of childhood – the hygienic childhood, the safe childhood, the brightly colored childhood in which everything, even bathtub articles, has been designed to please the childish mind' (Hohn 2012: 225–6).

Hohn's Plastic Pacific entails following the movements of the rubber ducky as flotsam – the way it connects disparate sites and interests, the object- and thing-stories it evokes, rather than the material itself (or the risks to life plastic poses, cf. ibid.: 3–4). The biological risks of plastic alluded to here are, in Hohn's account of the spill, backgrounded to the financial risks of flotsam and jetsam. The distinction between flotsam and jetsam is a legal one, one that narrativizes the events that precede the arrival of the objects on the shore. That is to say, a story is formed to trace the objects prior to their arrival, and to categorize them. Whereas flotsam references objects from wreckage or cargo floating on or washed up from the sea, jetsam is used to designate objects that have been thrown overboard to lighten a vessel, in particular when the vessel encounters difficulties at sea.[16] The event of arrival encompasses a (presumed) event of departure – in the case of 'Hohn's' rubber duckies, two containers tipping of the deck of *Ever Laurel* in the stormy seas of 10 January 1992 – and a trajectory or movement from the spatio-temporality of the departure to that of the arrival. The narrativization of the event of arrival is thus also a narrativization of what is not present: the objects (human, non-human) that wash up on the shore 'point' to themselves, their being 'here-now', and 'point' elsewhere, their having being 'there-then': 'elsewhere' and 'elsewhen'.

Such an account of the distinction between deliberately jettisoned objects (jetsam) and accidentally lost objects (flotsam) is a narrativization stressing the intention of an unseen agent. The agent is, in juristic accounts, necessarily human. However: agency before the law extends to incorporated institutions, that is, corporate *bodies*. Accountability for loss, as Hohn is at pains to stress throughout the book, is rendered moot when the risk is figured as *financial* risk to the corporate bodies involved, a nasty 'tradition' in European maritime disasters that encompasses the Zong massacre of 1781, when enslaved humans were thrown overboard by other humans, only in order to collect insurance. Financial risk, such stories suggest, recognizes that ethics do not extend to all *bodies* equally, and are not extended to the bodies of water (which must absorb the objects) or to the bodies of critters (which absorb the materials and toxicants), either. The law, and with it, specific narrativizations of agency, acts to privilege certain bodies and to foreground particular kinds of harm.

Plastic is entangled in all our lives. This entanglement reaches beyond that depicted in images of a rubber duck bobbing on the waves of the Pacific or the pervasive images of animals caught in plastic objects (e.g. plastic bags or six-pack holders). Thinking plastic as objects – broken and reconfigured – must give way to thinking plastic as material, as flows broaching object boundaries: trans-corporeal risks posed to bodies, human and non-human, rather than financial risks of loss of profit (as the flotsam narrative suggests). The material, objects broken and reconfigured, allows for comprehension in terms of horizontal, rhizomatic, plural relations in a manner that resists a more simple, singular subject-object relation.

---

[16] The legal distinction between flotsam and jetsam varies depending on local and national jurisdictions, with specific differences in the US American and British contexts, for instance (cf. Kenny and Hrusoff 1967).

Lipophilia and the risks of bath-time fun (rubber duck ii)

In their book *Slow Death by Rubber Duck: The Secret Danger of Everyday Things*, Rick Smith and Bruce Lourie mobilize the rubber duck as an 'innocuous household icon' (Smith and Lourie 2009: xvi) to undertake a project of self-exposure to the harms and risks of 'everyday' materials. The cover of their book depicts a plastic rubber duck that has been incorporated into a minimalistic diagram, with labels alerting the viewer to the presence of heavy metals (chromium and lead), other toxic elements such as bromine and chlorine, as well as phthalates (the substances that make plastic 'plasticky'). The book traces their investigations into the relations between environmental and health issues, following a model familiar from Morgan Spurlock's film *Supersize Me* or some of Michael Moore's works (cf. ibid.: 2). In the accounts of *Slow Death by Rubber Duck*, Smith and Lourie expose themselves to seven chemicals: phthalates, teflon, brominated flame retardants, mercury, triclosan, quicksilver and bisphenol A. Of the seven, only two are not related to petrochemicals directly (mercury[17] and quicksilver). The rubber duck is not mobilized in this account as a symbol of the gyres of the oceans, or pelagic waste, nor as a symbol of the externalization of (financial) risk (as in the discussion of the legal distinctions between flotsam and jetsam earlier). However, its iconicity suggests that it can forge such relations and thus connect back to the concerns of the Plastic Pacific.

The rubber duck of the title and cover page image are mobilized to specific effect. Smith and Lourie explain: 'As one of the most charismatic phthalates sources around, the yellow icon, beloved by Sesame Street alumni everywhere, took centre stage in the ongoing U.S. phthalates debate' (ibid.: 57). Global production of phthalates, they suggest, is in excess of 8.1 billion kilograms a year (cf. ibid.: 34) and is used as 'plasticizers' (over 60 per cent for plasticizing vinyl (e.g. in PVC)), as well as in personal care products (specifically to lubricate other substances and to prolong fragrance) (cf. ibid.: 35). Consumer groups and lobbyists in California, and later on the US national stage, engaged in what they dub the 'Rubber Duck Wars', a debate about the toxicity and harm of phthalates in plastics, which forms the central concern of their second chapter. In their discussion of the debate, in particular with regard to the progress of bills in these jurisdictions, two photographs are included: both include rubber ducks. One

---

[17] And yet, the Minamata Mercury disaster was linked to the manufacture of polyvinyl plastic, as the industrial waste from a factory was being dumped directly into the bay where it was being absorbed and biomagnified through the food chain back up to the human species.

shows a hand-drawn sign with 'No Yucky in my [picture of rubber duck]', the other a button with the slogan 'Save the Rubber Duckies' (cf. ibid.: 58). In Smith and Lourie's book, the rubber duck acts as a placeholder for an exploration of phthalates' impacts on human health. It also works to amass sympathy by its foregrounding of the symbolic weight of the rubber duck.

The weight of the choice of the rubber duck in this respect is to emphasize 'our children' and, with it, an attempt to evoke an ethical relationship towards future (possibly genetically related) generations. A common trope, the appeal to children is widespread, not only in the appeal to 'future generations' by environmentalists, but also omnipresent in advertising (think of insurance, banking or house loan advertising, for instance). The trope of appealing to children is problematic. As Naomi Klein relates in 'The Right to Regenerate' (in *This Changes Everything*), her personal account of fertility issues led to resistance to these kinds of invocations:

> If I was going through a particularly difficult infertility episode, just showing up to a gathering of environmentalists could be an emotional minefield. The worst part were the ceaseless invocations of our responsibilities to 'our children' and 'our grandchildren'. ... [W]here did that leave those of us who did not, or could not, have children? Was it even possible to be a real environmentalist if you didn't have kids? (Klein 2014: 423)[18]

Queer thinkers also have stressed that the problematic evocation of children in environmental discourse sometimes is an emblem not so much of (but also) human reproduction, but more broadly as a sign of the reproduction of particular ways of being in the world. Heather Davis, for instance, draws on Donna Haraway, Nicole Seymour and Rebekah Sheldon to argue that 'the Child [capital C] becomes the stand-in for a certain mode and expression of inheritance that often serves to uphold or buttress current social and power formations. ... However, what these discourses are often seeking to protect is not the health of any future child but rather the maintenance of a particular way of life' (Davis 2022: 91).[19] In particular, Davis's mobilization of Sheldon's articulation of the child as 'a kind of retronaut, a piece of the future lodged in and under the controlling influence of the present' (Sheldon, as quoted in Davis 2022: 91) dovetails with the idea of the future artefact I articulate in this project.

---

[18] Sometimes, the invocation of children works in a diametrically opposed manner: I have witnessed well-known academics arguing that it is impossible to be an environmentalist if you *have* children.
[19] Liboiron, Tironi and Calvillo argue, more forcefully: 'More than just the contravention of an established order within a system, toxic harm can be understood as the contravention of order at one scale and the reproduction of order at another' (2018: 335).

But, children are also not just tomorrow's adults. Children, as Smith and Lourie argue, interact with their environments in a different manner to adults, particularly in their propensity to the tactile (often in conjunction with their mouths). As they put it, 'by virtue of being closer to the dust bunnies, licking their fingers relentlessly and chewing on phthalate-containing items that they shouldn't be putting in their mouths, ... kids are sucking in more of this stuff' (Smith and Lourie 2009: 38). And, children are particularly vulnerable to all sorts of toxicants, especially in particular phases of development (cf. e.g. Smith and Lourie 2009: 204–6 and Klein 2014: 427[20]). Children's vulnerability towards toxins patterns differently to that of adults: They are not to be considered 'mini-adults' whose bodies engage with and react to (toxicants in) the environment on a scale dictated only by size.

The rubber ducky works to encapsulate relations otherwise disparate and difficult to visualize, at the same time as mobilizing a(n assumed) concern for the fate of children. The compound that attracted much of the attention in the 'Rubber Duck Wars', and which circulates by its very absence even now after being banned, is bisphenol A.[21] Smith and Lourie use a list – another concatenation of plastic – and a table in their attempts to approximate the issue. The list is an extensive inventory of household products which harbour BPA:

> A typical house is chock full of BPA. Polycarbonate plastic is used to make CDs and DVDs, water bottles, drinking glasses, kitchen appliances and utensils, eyeglass lenses ..., bottled water carboys (the big water jugs used in office water coolers), hockey helmet visors, baby bottles, medical supplies and the faces of my laptop and Blackberry. Polycarbonate is also extensively used in cars and trucks for things like headlights, and it's right there in my kids' toy bin .... They're also commonly found in dental filling materials, protective coatings around wire and piping and what is likely the primary avenue of exposure for most people: the interior lining of virtually every tin can found in every home and grocery store. (Smith and Lourie 2009: 228)

The effects of BPA tabulated in their book include permanent changes to genital tract, disposition to cancer in breast cells, interference in brain structure and function (also reversal of sex differences thereof), as well as other hormonal

---

[20] Klein notes, citing Sandra Steingraber, 'Until 1990, ... the reference dose for radiation exposure was based on a hypothetical 5'7" [approx. 170 cm] tall white man who weighed 157 pounds [approx. 71 kg]' (Klein 2014: 427). Klein's point here is to illustrate the ways that children and unborn foetuses are left out of many studies. Beyond that, the reference 'norm' is also obviously gendered and racialized.

[21] How does BPA still circulate despite being banned? As discourse, through packaging, for example, of baby bottles, that declares, prominently, 'BPA-free'.

changes (cf. ibid.: 234–5). Other health concerns linked to BPA include decline in semen quality in men, early onset of puberty in girls, ADHD and other neurobehavioural problems, as well as metabolic disorders, like diabetes and obesity (cf. ibid.: 237). One of the key issues of BPA is that these effects take place at particularly low concentrations, contradicting the previous logic that had suggested that larger doses would be more dangerous and vice versa. As BPA mimics hormones, it makes sense that our bodies are sensitive to small doses; indeed, BPA acts completely differently on our bodies at small doses than it does at high doses. And it is omnipresent: BPA is 'rapidly metabolized by the human body' (ibid.: 228) so the fact that a vast majority of US Americans test positive (Smith and Lourie cite 93 per cent, cf. ibid.: 227) meaning that these bodies are continually exposed, or are 'marinating in BPA every day' (ibid.: 228). Another concern is that of scale: 'constant exposure to industrial chemicals' as Max Liboiron, Manuel Tironi and Nerea Calvillo stress, 'is paired with the fact that many of these chemicals persist in a geological time frame that exceeds the timescale of the human species, meaning chemical legacies will characterize the planet for both immediate and distant futures' (Liboiron, Tironi and Calvillo 2018, 332). And, as Max Liboiron notes in *Pollution is Colonialism*, 'Scale is not about relative size. Scale is about what relationships matter within a particular context' (Liboiron 2021: 84).

BPA was first synthesized in 1891, and its hormone-disrupting properties discovered in the 1930s, so whilst it is not a recent 'discovery', it has invented particular, toxic, relations. The harnessing of the image of the rubber duck – on the cover, in the title, and in the account of the 'Rubber Duck Wars' – works to suggest a toxic agency present in the seemingly innocent objects of (childhood) surroundings. In their 'self-experimentation', further, Smith and Lourie work through the issues of expert and layperson accounts of risk and attempt to develop a compelling narrative of these risks. Marco Armiero might call it a 'guerilla narrative' of the Wasteocene, where the idea 'is that toxic storytelling does not only uncover the traces of toxicity, exposing the injustice embedded into the Wasteocene, but it also frees an antagonist narrative which can potentially transform contaminated communities into resisting communities' (Armiero 2021: 24).

In 'Toxins, Drugs, and Global Systems: Risk and Narrative in the Contemporary Novel', Ursula K. Heise notes a shift in the conceptualization of risk that recognizes that lay opinions are not readily rectified by explanation or correction, but that such opinions or assessments are themselves constitutive of risk: 'Mediating processes and institutions … shape the social experience of risk and magnifiy or minimize it', leading to the question of 'not only whose risk assessments are more realistic but also what criteria should be used to gauge degrees of risk in the first place' (Heise 2002: 761). Heise accordingly suggests that *narrative* analysis might be useful for eliciting the manifestation of risk in literary and nonliterary writings (cf. ibid.: 762). Drawing on Charles Perrow's work on 'systems accidents', Heise notes that the risk emerging from technological failures is complex, in that it is not only rarely traceable to singular causes (and, hence, rarely 'fixable' or preventable by focussing on particular components), but rather contingent on the interactions of systems, where minor failures may lead to major accidents: contingencies, but also entanglements.

The 'accidental' in Heise connects the narrativization of risk, Armiero's guerrilla narrative, with that which is implicit in the distinction between flotsam and jetsam articulated earlier: that stories matter. The story of Hohn's rubber duckies, thus, begins earlier than January 1992. It could be said to begin in the factory of its production (which is where Carle begins his story). It could also be said to begin much earlier: with demand, oftentimes manufactured by manufacturers. Such stories arise in an abundancies of 'elsewheres', entangling our lives in multitudes of ways: with the plastic relations that gave rise to the inventory of its manifestations, its invention, with the realization that plastic relations stretch beyond the across the temporal dimensions implicit in its material, and that the material itself forges (toxic) relations that exceed its manifestation as object.

At the same time, in many forms and objects, plastic serves to isolate risk. It is omnipresent in the medical sector, in the form of disposables that minimalize the risk of transferring infection or disease. Uses of plastic comprise intravenous bags and tubes, intubations, bottles, syringes, gloves for medical practitioners and so on; it functions as a barrier, demarcating a particular border between self and other, between body and (potential) containment. Plastic food wrapping functions in a similar manner, holding off decay and decomposition, manifesting a border between nourishment (future-body) and perishability. These kinds of practices will form the primary concern of Chapter 4. In *Slow Death by Rubber Duck*, this 'everyday' plastic is shown to be both innocuous and insidious. Their subtitle 'Secret Danger of Everyday Things' works perhaps as a ghost story, where the ghosts are of the future or of elsewhere, held at a distance, but, at the same time, present in the home: a spatio-temporal othering rendered internalized.

Seas of plastic/plastic oceans

Charles Moore's *Plastic Oceans*, written with Cassandra Phillips, offers a personalized account of Moore's encounters with the plastic of the oceans, specifically the Pacific. Classified on the back cover as 'Environmental Science', the book 'proper' starts with a startling simile: 'The ocean looks like glossy blue cellophane, like a pond on a summer day' (Moore and Phillips 2012: 1). Later, however, cellophane is 'considered a bridge material between plastics and paper' (ibid.: 146). And, the ocean pondered at the outset is shown to be more like cellophane than initially imagined: 'I begin to notice that this smooth "painted ocean" seems to be – how best to put it? – littered. Here and there, odd bits and flakes speckle the ocean surface. ... A bottle here, a bottle cap there, scraps of plastic film, fragments of rope or fishing net, broken-down bits of former things' (ibid.: 3). The illusion of a plastic material bridging plastics and paper is revealed as *consisting of* plastic, bridging oceans and continents.

The narrative of *Plastic Oceans* is concerned with Moore's attempts to garner public and scientific attention for the plastic accumulating in the Plastic Pacific. The book accordingly narrates his many forays into academic scholarship and his frustration at the lack of attention given to the issue of plastic in the oceans. Attending the Fourth International Marine Debris Conference on Derelict Fishing Gear and the Ocean Environment held in Honolulu (the Pacific!) in August 2000, he recognizes that the other attendees are concerned with plastic as objects, not as material. The distinction is important for Moore, as it is for this chapter. As objects, Moore noticed, the other attendees are responding to concerns encoded in visual terms:

> They're [the attendees of the conference] all about massive clumps of plastic net, monofilament line, floats, and buoys. ... I feel like I'm peering out from a diving bell, exploring new underwater terrain. And from this vantage point, it's also clear that plastic, the material, is of far less concern than the derelict nets and other types of fishing gear made from it. (Moore and Phillips 2012: 158)

At the Honolulu conference, Moore attends a 'solutions workshop'. He notes: 'No one seems to want to go to the heart of the problem. It's the *plastic material itself* that is formed into almost every debris thing out there' (ibid.: 162, emphasis added). In line with the title of his book, then, Moore narrativizes his account in terms of oceans, both as sites of connectivity, and in the sense of 'overwhelming mass': He takes recourse to the stylistic device of the list – yet another concatenation of plastic – ostensibly taking the form of a collection of objects. Through the employment of many lists throughout the book, he evokes a profusion of plastic permeating our day-to-day lives. We are, it seems, swimming in it, even as we remain firmly on terrestrial grounds.

By referring to plastic as 'the solid phase of petroleum' (ibid.: 328), Moore can think of the pollution it causes as an oil spill 'that last centuries and mimics food while sponging up toxics' (ibid.: 329). I employ a similar move in the proposal of the 'future artefact' in the introduction to this book. The agency of plastic, here, is a function of its capacity to exert influence beyond its use for humans. The lipophilic ('fat-loving') characteristics of plastic mean that chemical components of some kinds of plastic, such as PCBs and PBDEs (polybrominated diphenyl ethers), are absorbed by animals – including humans. They also biomagnify; that is, they accumulate as they move up the food chain. Further, some kinds of plastic, for instance, pre-production pellets, adsorb toxicants (called persistent organic pollutants, POPs). It follows that plastic does not only accumulate in discrete forms, but it also attracts and exudes toxicants, pushing back against the ostensible boundaries of visible forms. We need to think of material, of plastic, in its stretchiness and in its condensed forms, implying entanglements both constitutive as well as destructive, and for the ways it is complicit in modernity's drive for development: 'Synthetic Evolution' (to quote the title of one of the chapters) is tautology as much as oxymoron.

The challenge Moore has set himself is to gain attention for a *material* rather than a *thing* or *assemblage of objects*. In this respect, as should be obvious by now, I find the organizational motif of the ocean auspicious. Despite Moore's conviction that he should be able to 'mobilize the masses' behind the mass of plastic as a problem in and of itself, and, further, as the story of the problems encountered at the conference in Honolulu reveal, he finds himself increasingly frustrated at garnering attention for the issue of plastic. Richard 'Ricky' Grigg, 'a champion surfer, Stanford graduate, and Scripps-trained University of Hawaii professor' (Moore and Phillips 2012: 116), points out to Moore that he needs to show harm (cf. ibid.: 118): 'I say a man-made synthetic is extremely likely to be entering the food chain: how can there be anything good about that?' Grigg responds by saying, 'Groups like the UN, the main international body dealing with non-sovereign ocean issues, don't care if the ocean is full of plastic. They will want hard evidence of harm before considering policy changes' (ibid.: 118). At least two insights crucial to the Plastic Pacific are evident here: firstly, that plastic is afforded an agency, and, secondly, the unspoken assumption that harm must be towards humans (and not, say, albatrosses). Some bodies, it seems, matter more than others.

In one vignette, Moore recounts how a group of Japanese scholars working on marine toxicity realize that plastic pellets were 'potential tools for assessing marine toxicity. How perversely fitting: a pollutant monitoring pollutants' (ibid.: 187). Nurdles, as pre-production plastic pellets are called, attract toxicity and thus, when consumed by wildlife (the Japanese group was focussed on seabirds), bio-accumulate.[22] This quality of plastics derives, notes Moore, from its origins as oil, as decomposed organic matter:

> It's in the nature of these chemicals, given their oil-based essence, to be drawn to like substances – fats, oils, lipids. All living creatures are composed of three basic ingredients: carbohydrates, proteins, and lipids. Thus, *we* attract and indeed harbor POPs [persistent organic pollutants], as do the creatures of the contaminated oceans, and as do, I was realizing, the processed hydrocarbons we know as plastics. (ibid.: 185, emphasis in original)

Plastic's toxicity is a direct function of its material origins, stretched to the relations that generate the oil from which it is made. For the considerations of plastic's toxicity in this conjunction, it is the relations that stretch across non-human scales of time – back to the age of the dinosaurs and beyond and also towards an (uncertain) future – that comprise the material and give rise to its future relations through its very 'oil-ness' that are crucial here.

A vital materialist account, further, allows for a consideration of the relations forged by (discarded) plastic (in the Pacific, and elsewhere) that enact changes on biological matter at the level of DNA in human beings and other non-human animals. Moore, for instance, notes the following:

> The most contaminated people ever biomonitored live in northwest Greenland. There, two girls are born for every boy. The mother's lipid PCB [polychlorinated biphenyl] level correlates with her likelihood of delivering a girl. Moreover, the Danish scientists who studied this community found that the babies tended to be born preterm and small, which often portends developmental and neurological

---

[22] Micro-plastics are categorized into two types: pre-production plastic resin pellets, or nurdles, typically less than 5 millimetres in diameter (primary micro-plastics), and the fragments that are formed from debris by weathering and other processes of degradation. The former, notes Koert van Mensvoort, are produced to aid the shipping of large quantities of plastic:

> Over 100 billion kilograms of nurdles are shipped each year as raw material to processing plants, to be heated up, stretched and moulded into the plastic products and packaging so familiar to us. … The most common source for ocean-bound nurdles is industrial spillage from trucks and container ships. Because nurdles are so small, they are hard to contain, slipping away effortlessly from containers into waterways or directly into the ocean. You can find nurdles on virtually every beach, hence their nickname: mermaids' tears. (van Mensvoort 2011)

Slippage, spillage.

problems. This is a sign of endocrine disruption, of genetic blueprints being altered by subverted hormonal messaging during gestation. (ibid.: 251)

These genetic alterations are the result of the bio-accumulation of POPS, persistent organic pollutants that are 'passed up' the food chain through the consumption of organic fatty matter. The Inuit of Greenland, Moore points out (in a rather cursory phrasing), are 'polluted by their "natural", traditional diet, a fact they are aware of and unhappy about' (ibid.: 252), that is, as maritime carnivores, they are eating accumulated toxicity. Raj Patel and Jason W. Moore have argued that capitalists are 'happy to view the ocean as both storage facility for the seafood we have yet to catch and sinkhole for the detritus we produce on land' (Patel and Moore 2020: 23). Here, the distinction hardly holds; Charles Moore ends his book with the assertion that 'our bodies have been polluted as much as the oceans have been' (ibid.: 357).

In common parlance, there is a tendency to use 'toxic' as a synonym for 'polluted'. There are two key problems to such a conflation. Firstly, toxicity does not only suggest the 'state of being polluted' but also suggests the 'potentiality for polluting'. So if something is toxic, it has not only transferred into that state (ostensibly from some kind of 'clean' state), it also bears the weight of potentially contaminating those things around it, pushing them into this state with it. Toxicity provides a case study, then, for the agential capacity of non-human actants. More than an *a priori* agency, the agency of toxicity emerges through relations. Its agency is expressed through relations with other bodies: What might be toxic to me, might not be toxic for a dog (or vice versa, for instance: chocolate), or indeed for another human whose body has different vulnerabilities.

Secondly, toxicity has no apparent index. Where pollution has an (over-)abundance of indices, toxicity renders its presence in more malevolent means. Pollution's indices are often visual – as not only in the ubiquitous plastic bag or abandoned face mask, but also the smoke pouring out of factories' stacks and exhaust pipes, sludge on river beds, landfill sites and oil slicks, to name a just few examples. Pollution also produces olfactory indices or smells. The smoke might have a pungent smell, rendered bitter on the tongue and the sludge might exude acidic or bitter vapours. For Marco Armiero, this immediate sense is integral to many of the toxic autobiographies he has collected in his work as a historian (cf. Armiero 2021: 25).

Toxicity's indexes take time to materialize: the dying trees of the Black Forest of biology text books of the 1990s, for instance, or the manifestation of mutated genes in all manner of biota following nuclear fall-out in the Nevada, Chernobyl or Fukushima regions (and of nuclear extraction regimes, for instance, in Gabon, cf. Hecht 2018). Rob Nixon, in *Slow Violence and the Environmentalism of the Poor*, suggests that the toxicity engenders metamorphoses of social justice and 'collective relationship to apocalyptic time' (Nixon 2011: 58). The scale, and global reach, of toxic materialities is indicative of material manifestations of the Anthropocene. Plastic practices render relations of the Plastic Pacific risky, harmful, even toxic.

Towards an archipelagic Plastic Pacific

To concentrate on this line of argument is, however, to replicate some aspects of the thinking that have produced the (environmental, health) concerns in the first place: the Pacific is not simply ' "vulnerable", "disappearing", "sinking", and "endangered", and thus in need of "saving" ' (Perez 2021: 430). The Pacific might figure as a sink for pollution, as a playground for exploration, or as an otherwise exoticized space, but it is also, for some people, home. Following from Craig Santos Perez's call to resist reductionist accounts of the Pacific, and the livelihoods of the people who inhabit the Pacific (both literally and metaphorically, as is the case of Pacific Islanders who have migrated to other places, cf. ibid.), I now turn to Perez's own poem 'Age of Plastic' included in his 2020 volume of poems *Habitat Threshold* (Perez 2020: 11).[23]

In this poem, we find many of the strains of this chapter woven together: references to childhood and children, to medical practices, to waste, to the Plastic gyre, even to birds (who with whales, plankton and shrimp '*confuse plastic / for food*' (ibid., emphasis in original). Condensed into a one-page poem that is literally littered with plastic (the word is included eighteen times), plastic figures forcibly as a material, dwelling on and with its materiality.

Paired with a graph on the facing page (verso) entitled 'We are not drowning …',[24] the poem is also accompanied by a quotation from Roland Barthes's musings on plastic in *Mythologies*. Typographically, Perez makes use of two

---

[23] The version included in 'The Margins' with its 'Countertranslation' by Divya Victor is quite different. The online version does not use the bold typeface, is shorter, does not have the epigraph from Barthes and also ends on a more sarcastic note (the lines 'So that she, too, will inherit / "a great future"' are missing from the version in *Habitat Threshold*). Victor's commentary dwells on the terraqueous qualities of the process of translation and also evokes his daughter: links, here, stretching to rubber duckies, as well as into the considerations of Chapter 4, in particular Jody A. Roberts's essay that I discuss there.
[24] See Martin Premoli (2021) for a close analysis of Perez's graph, which adopts a graph from the Intergovernmental Panel on Climate Change (IPCC) report.

shifts: between italics and regular typeface, and between bold and regular (some words use both italics and bold). Shifts between the regular and italics trace personal observations and evocations of the quotidian in regular font, and facts, statements and questions about the role of plastic in italics; this marking serves to foreground the impersonal – assumptions, assertions, generalizations – as the deviation from the norm. Bold typeface is reserved – in the poem and epigraph alike – for the word plastic (even as it comprises a root word, as in '**plastic**ized' from the epigraph).

Like the other writers concerned with plastic, Perez's poem elicits many different encounters with the material: 'Plastic' is scattered throughout the quotidian and personal (regular font), as well as the declarative and impersonal (italics); the stress of the bold shows how the omnipresence of plastic accrues at multiple scales. In 'Age of Plastic', plastic is, amongst other things, the material of medical devices (see also Chapter 4), toxicant, receptacle, technological wonder, shaped such that it can soothe a child (as pacifier, or 'dummy' as I would call it), mass material, agential, SF wonder, and durable. These multiple materializations reverberate with the evocation of the 'Plastic Age', through Perez's title, evoked in Yarsley and Couzen's 1941 book *Plastics* (see also 'Introduction'). This piece imagined a world enabled, and saturated, by plastic. There is also an echo of the iconic photograph from *Life* magazine's 1955 cover entitled 'Throwaway Living', depicting a black-and-white image depicting a child, woman and man throwing all matter of plastic goods up into the air (in what we imagine must be a celebration, but one that jars from today's perspective, see also above).

However, in contrast to other writers who use what could be called the 'concatenation of plastic' trope, the repeated emphasis on the word 'plastic' renders this a material connection. The repetition and stress afforded the term 'plastic' means it shifts subtly, transcending (whilst not forgoing) the manifestations as object; it is, again, present both at the local scale of individual lives and at a global scale, evoked by the shifts in typeface. With this stress of the material, the reader is, I want to argue, invited to think through all the different manifestations of plastic, and to link these different manifestations together, broaching these seemingly disparate scales to afford a moment of pause for the saturation of plastic in all our worlds.

Thinking the 'Great Pacific Garbage Patch' with archipelagic thought entails considering difference as a function of degree and scale – of distance and time – rather than as an absolute.[25] The constitutive difference of land and water

---

[25] Archipelagic relations, Stratford et al. write in 'Envisioning the Archipelago', 'may provide novel opportunities to unsettle certain tropes: singularity, isolation, dependency and peripherality;

(together with land and land, both island/mainland and island/island) inherent in this rethinking of patterns of identity and difference emphasizes cultural (and historical, social, political, ecological, etc.) patternings of island places and spaces. These islands of difference are placed in relation with each other, and with the flows that connect them, through the archipelago. The 'Plastic Pacific' emphasizes the shifting accretion of material rather than the static accumulation of objects.

In Perez's poem, the entanglements of plastic are no longer abstract or 'elsewhere', as might be the case in some of the other texts I analysed earlier: As a CHamoru poet from Guåhan residing in Hawai'i, Perez's concern for the Plastic Pacific is tied to places *of* the Pacific. These ties are not 'grounded' in biography, however, as the verb 'to ground' suggests a privileging of terrestrial inhabitations. Perez's academic work on archipelagic thought cautions against such (metaphorical) privileging of landed relations, cautioning against the 'subordination of island spaces to dominant continental spaces' (Perez 2017: 100).

Habitat Threshold includes many more poems that address other environmental issues of and in the Pacific. The volume also encompasses adaptations, or 'recycling' as the poems are subtitled: Dr Seuss's 'One Fish, Two Fish, Red Fish, Blue Fish' transforms (but doesn't downcycle, as is usually the case with material matters) into 'One Fish, Two Fish, Plastics, Dead Fish'; Pablo Neruda's 'Sonnet XVII' becomes 'Love in a Time of Climate Change' (replete with reverberations of Gabriel García Marquez's novel, pathologizing Climate Change as it replaces Cholera) and poems/songs by Wallace Stevens, William Carlos Williams, Irving Berlin, Allen Ginsberg and Maggie Smith also are 'recycled' in the volume. Other thinkers find homage in 'Praise Song for Oceania', as the Acknowledgments recognize the thought and poetry of an array of ocean(ic) writers. The graphs, too, comprise reconfigured and reorganized ideas. The cover of the edition I analyse includes a photo depicting Perez's father and daughter at the beach as its main image; the image at the top of the page shows ice (possibly glaciers), the image at the bottom shows fire. The back of the book is mostly covered by a box of endorsements by other authors; the background, however, depicts various lurid objects against a light, beige background: pieces of plastic against beach sand.

---

perhaps even islandness and insularity' constituted in 'a form of counter-mapping' that manifests in 'different combinations of affect, materiality, performance, things' (Stratford et al. 2011: 114). And, as Paul Carter argues, 'The archipelago is not only a collection of islands, it is the labyrinth of passages between them[.] ... [T]his has an epistemological analogy, for an archipelagic sense of places rejects the totalizing vision of imperial (or continental) geography as an impossibility' (Carter 2014: 185).

Elizabeth DeLoughrey has invoked 'tidal dialectics', following Kamau Brathwaite's 'tidalectics', as a 'methodological tool that foregrounds how a dynamic model of geography can elucidate island history and cultural production' and 'resists the synthesizing telos of Hegel's dialectic by drawing from a cyclical model, invoking the continual movement and rhythm of the ocean' (DeLoughrey 2007: 2). If the persistent characterization of the island as isolated is a function of the erasure of technologies of mobility, then the movement of plastic in the gyre works to counter this impression. Indeed, archipelagic thought insists on avoiding thinking in terms of absolute difference, stressing the connections as well as the separations (cf. also e.g. DeLoughrey 2001, Roberts and Stephens 2013, Stratford et al. 2011, Baldacchino and Clark 2013). When plastic objects are considered in terms of their material, a different, non-anthropocentric agency emerges: plastic becomes, well, more plastic.

Alice Te Punga Somerville reckons with the 'Great Pacific Garbage Patch' as a 'manufactured' archipelago (drawing on the work of Jonathan Pugh), aligning it with all kinds of objects and interventions of human design. 'Like any archipelago', Te Punga Somerville writes,

> The great Pacific garbage patch is made up of constituent parts whose medium both of connection and disconnection is the ocean. Like any archipelago, there are general currents and tides that affect the whole larger entity ..., but there are also extremely diverse experiences and entities within that system. (Te Punga Somerville 2017: 323)

Te Punga Somerville also observes that the outsider comes to the GPGP and at first only encounters – even perhaps just *sees* – the disparate pieces; it is only on further consideration and examination that they are able to recognize the extent to which these pieces are connected (cf. ibid.). As shown earlier, the toxicity of plastic as material cannot be decoupled from its manifestations in objects, and the (often fleeting) use of plastic in terrestrial practices cannot be decoupled from its presence in the oceans. Toxic relations demand a rethinking of agency (beyond its traditional ascription only to humans, or perhaps to mammals) and a rethinking of the borders around biota, species, individuals and the material manifestations of objects. In the Plastic Pacific, the barriers between terrestrial and marine, between object and material, and between the external(ized) and the internal(ized) are rendered porous: Or are rendered chains, like the polymers that plastic itself is made of, as *chains of islands*.

3

Megadam materialities

Introduction: Megadam materialities

In this chapter, water comes into force as an element whose flows can be arrested and released as a means of creating power. In Chapter 2, water was the element that provided the flows and shifts of the main site: pelagic plastic bobbing on and whirling just below the surface of the ocean, and porous plastic, materializing as toxic threat, shifting with the tides. In this chapter, the concrete power of megadams[1] comes into being through the shapes it takes: through the structures of the dams themselves, through the infrastructures they assemble and through the responses they generate.

The power of megadams is generated as it is resisted; as electrical power – the kind that 'powers' electronic devices, the laptop I write on, the lamp that lights my room, the refrigerator that prolongs the viability of my food – and as discursive power, the forms through which political power asserts, maintains and folds in on itself. The concrete forms of megadams have a material vitality (Bennett 2010), an 'agential realism' (Barad 2007), a potential to engender politics: political tensions and aesthetic renditions. Megadams are, crucially, infrastructural: In their introduction to the special issue of *Modern Fiction Studies* on 'Infrastructuralism', Michael Rubenstein, Bruce Robbins and Sophia Beal suggest that infrastructure 'is supposed to go unnoticed when it works; destroying it is simply one strategy for making it appear … converting it into

---

[1] Regarding the term 'megadam': categorizing a dam in terms of its size is difficult, as Kate Showers argues, given that there are many measurable characteristics, for example height of wall, volume of water stored, surface level of water stored and so on. These variables have no consistent relationships, as dams in deep gorges, for instance, will have small surface levels despite large volumes of water and/or heights of walls. Further, there is no consistency in the literature on dams regarding terminology (cf. Showers 2011: 1658–9). Showers also notes that the definitions vary with respect to (disciplinary) perspective: Engineers, she argues, will be primarily concerned with wall height, whereas 'ecologists and social scientists are more concerned about the size of the reservoir created' (ibid.: 1660). I use the term 'megadam' rather indiscriminately to mean 'very big indeed', to suggest impact.

a proper spectacle' (Rubenstein, Robbins and Beal 2015: 576). The idea of the spectacle of the seemingly ordinary comes into its own in this chapter through the multiple explosions of the Glen Canyon Dam. Concrete's ubiquity as a material and its role terraforming environments with local manifestations and global effects – not least as an index of development – is also explored. Amongst other functions, a sense of the 'poetics of infrastructure' articulated by Brian Larkin becomes evident here: 'Infrastructures are the means by which a state proffers these representations to its citizens and asks them to take those representations as social facts. It creates a politics of "as if"' (Larkin 2013: 335).

The consideration of megadams is, like the previous chapter, connected to environmental concerns: The planned hydro-electric project on the Franklin River in Tasmania occasioned protest that laid the foundations for the Green Party of Australia. The forerunner for this political party – the United Tasmania Group – is considered the first to campaign on a predominantly environmental platform (cf. e.g. McCann 2012). Many of the texts I consider in this chapter lodge such environmental justice concerns firmly in the concrete forms of megadams: Leslie Marmon Silko's *Almanac of the Dead*, Edward Abbey's *The Monkey Wrench Gang*, the World Commission on Dams report from 2000, two documentaries on the Glen Canyon Dam (one by EarthFirst!, the other commissioned by the United States Department of the Interior Bureau of Reclamation), Thomas King's (2012) collection *The Inconvenient Indian: A Curious Account of Native People in North America* as well as a number of essays by Arundhati Roy.[2] The capacity of megadams to produce power, and to engender political tensions (through aesthetic responses as well as policy documentation), is a key concern of this chapter.

This ostensibly eclectic set of texts reflects a patterning of megadam materialities through the local and global. This basic assumption, outlined in more detail in my introduction to this book, draws attention to the materials that comprise the materiality of concrete, that is, the sand and stones (aggregate), the cement (binder), reinforced steel (support) and water. Two of these are sourced locally – the aggregate and water – whereas the other two are globalized products – the cement and reinforced steel. These material dimensions reverberate with the local and global dimensions and effects of megadams. As Dogan Altinbek observes, 'Large dams are generally justified

---

[2] This list of Anglophone novels that address environmental justice in conjunction with megadams can be supplemented by Linda Hogan's novel *Solar Storms* (first published 1994, cf. also the introduction to this book) and Thomas King's novel *Green Grass, Running Water* (1993), as well as Anne Michaels's novel *The Winter Vault* (2009) and Louise Penny's *How The Light Gets In* (2013).

by regional and/or national macro-economic benefits, while their physical impacts are locally concentrated' (Altinbek 2002: 11). The environmental and social impacts – the more negative effects – of dams tend to be localized, whereas the benefits of the energy thus harnessed are often transposed elsewhere. The displacement of energy, and its concomitant mappings of power, also stress a patterning of localized and, if not globalized, then trans-regional, consequences of megadams. These map unevenly through the globe, producing powerful tensions and resistance. The attempts to harness the movement of water to produce electricity, one form of power, give rise to other power of other kinds.

The World Commission on Dams is explicit about the effects of dams: 'The debate about dams' it is argued 'is a debate about the very meaning, purpose and pathways for achieving development' (WCD 2000: xxxiii). Development is a tricky term, loaded with all sorts of baggage. Here, it is (partly) evoked to act as a chiffre for projects of modernity, in a way that draws on (postcolonial) critique formulated under the heading of alternative modernities (cf. Gaonkar 1999, Taylor 2001; see also the introduction to this book). It is the sense of development that is evoked by Robert Young, when he offers the following description of what postcolonial thought 'pushes back on':

> Colonial and imperial rule was legitimized by anthropological theories which increasingly portrayed the peoples of the colonized world as inferior, childlike, or feminine, incapable of looking after themselves (despite having done so perfectly well for millennia) and requiring the paternal rule of the west for their own best interests (today they are deemed to require 'development'). (Young 2003: 2)

Development, in an analysis of material cultural studies, cannot 'just' be financial aid: It must also reference the material processes through which states (i.e. political, material, even psychological) shift and change.

Concrete megadams, finally, are also to be recognized as indices. They have some kind of 'sensory feature' which links or correlates the sign with what it represents. As indices of development, megadams produce power and (re)produce power divisions. Such projects not only generate icons and manifestations of development through power in its various forms, but also through assimilation processes effected through displacement. These dimensions and tensions are to be considered in terms of layering, of a multiplicity of engagements. Dams are constructed in space and as a consequence of specific conceptions of space, in particular those that form relations towards some people and some things as resources ('extractive'), and further understood as one domain in and through which politics are manifest. The concreteness of concrete, accordingly, is key to this inquiry, and the repercussions of this emphasis are traced through various written and audiovisual texts. By affording attention to the material outcomes and the material processes that generate these outcomes, entanglement emerges as a political moment of ethical response to cultural texts.

The stuffs of our world – things, materials, companions, ideas – are not inert: form and function blend into each other; manifestation and potentialities shift as the materiality of the concrete dams pushes through its various forms and into imaginaries. Megadams assemble resistance (to dams, to power) at the same time as they generate power. Crucially, again, these effects accrue unevenly. In what follows, power is thus evoked as a deliberate coalescence of different kinds of power – including electrical power, political power, hegemonic power, resistant/subversive power – generated by and through the dams.

Vibrating power, indexical functions: On the shifting materialities of megadams

Megadams generate power through or by shifting materialities. They vibrate.[3] They are vibrant with power, in the best sense of Jane Bennett's *Vibrant Matter*. Recognizing megadams as vibrant matter and eliciting the entanglements of dams and power means entertaining a notion of vitality as 'the capacity of things – edibles, commodities, storms, metals – not only to impede or block the will and designs of human but also to act as quasi agents or forces with trajectories, propensities, and tendencies of their own' (Bennett 2010: viii). Noting, with Spinoza, that bodies are social in the sense that 'each is, by its very nature as a body, continuously affecting and being affected by other bodies' (ibid.: 21), Bennett's concern is for what she calls vital materialism, that is, how 'bodies enhance their power *in* or *as a heterogeneous assemblage*' (ibid.: 23, emphasis in original). Agency, here, is not localized with human or humanized bodies, but is distributed (cf. ibid.).

---

[3] This was suggested at a network meeting for the DFG-Network 'Environmental Crisis and the Transnational Imagination' held in February 2016 at the Rachel Carson Center in Munich. I am thankful to the members of the network, as to the contributors and responders at another workshop on 'Literary Environments' held at the University of Düsseldorf, and also the attentive listeners of a talk held within the University of Erfurt's 'Texte.Zeichen.Media' programme for their comments and suggestions on the various permutations of early work on this chapter throughout early 2016.

Figure 3.1 'Boulder Dam, 1942' by Ansel Adams. Photograph. Ansel Adams Photographs of National Parks and Monuments series, made available by the US National Archives (519840).

Power requires infrastructure. For electrical power, this infrastructure manifests in one form as high-voltage powerlines, as photographed, for instance, by Ansel Adams as part of a series commissioned by the United States Government, from which his iconic Hoover Dam photographs also emerge.

Ansel Adams's photographs of the various megadams of the US American West capture a sense of awe for the landscapes depicted. The photographer's oeuvre is usually considered in terms of his iconic depictions of wilderness, for example, the iconic 1942 photograph 'The Tetons – Snake River'.[4] Adams's depictions of architectural forms of the American (South-)West, in particular Pueblo cultural forms, also use the stark black-and-white format.

The megadams he photographs, for example, those in the Hoover Dam and the Boulder Dam series, are both subject of and subject to the images. Some images 'capture' the dams in the centre of the image, drawing the viewer's attention to their monolithic instalment in landscapes of the US American West. The 'Boulder Dam, 1942' photograph (Figure 3.1) allocates over a full two-thirds of the image to a depiction of the concrete surface, the contrast between dam and environment rendered stark through the exposure of the black and white image. The surface of the dam becomes a site of inscription in some of the other texts I discuss later, including the final image analysis of the last section of this chapter.

Other images by Adams show the infrastructural contexts of these dams, for instance, by shifting the focus to centre on the power plants or power lines. The 'Boulder Dam, 1941' photograph by Ansel Adams included as Figure 3.2 depicts infrastructural insertions into, and contrasted with, the landscape. Adams's photographs seem almost overexposed in their attempt to render the contrast as substantially as possible without losing any of the visual acuity of the subject, or *substance*. The image included here depicts the actual dam wall, half in shadows, half overexposed in the sunshine as well as a lopsided power line tower in the foreground. With a tilted pole, and exposure problems in rendering the contrast of light shed on the dam itself, this image shows a landscape 'out of tilt'.

At this intersection of culture and nature, or as an image of natureculture (cf. Haraway 2003; Latour 1993), Adams's images of megadam projects effect a contextualization, a framing of the incursions of infrastructural projects within the landscapes they occur. For these infrastructures are structures and networks alike, subject and object, foreground and background, sceptres of power. If, as Jane Bennett suggests, 'materiality is a rubric that tends to horizontalize' (Bennett 2010: 112), we might also look to the ways in which horizontal materials are imagined, and, importantly, the politics of such imaginations: intentions,

---

[4] This image was available online as part of the US National Archives at the time of writing: https://catalog.archives.gov/id/519904. It depicts a 'snaking' river, reflecting sunlight on its surface, with snow-topped mountains in the background.

Figure 3.2 'Boulder Dam, 1941' by Ansel Adams. Photograph. Ansel Adams Photographs of National Parks and Monuments series, made available by the US National Archives (519843).

enablings, resistances and patternings, with their impacts on specific livelihoods. Such considerations must redress the political dimensions of the way in which materiality 'horizontalizes', that is, to point out the ways in which such 'horizontalities' are not, necessarily or always, unstratified. If this is a grid that moves beyond structural restraints, or infrastructural holds, then it does so, coalescing and stretching, in uneven ways, across time and space.

Megadams as development

The generation of power by megadams is not restricted to the hydroelectric capacities of dams, although this is the focus of this chapter. Energy is also generated through dams by means of irrigation. Water won through irrigation is harnessed for agriculture: grains, legumes, fruits and vegetables grown with irrigation, and the water and food as that becomes fodder for animals farmed for consumption (by humans, or companion species), become (condensed) energy, measured in calories rather than watts. Power, through the generation of energy harnessed in these multiple ways, is thus crucial for thinking through megadams; megadams are crucial to the production of power.

Power is also resisted. The materiality of dams itself shapes the generation of power: by holding back water and by shaping landscapes through irrigation and inundation. Such shaping is evidenced in material forms through the monolithic megadam as an index of development in policy and other (aesthetic) cultural artefacts and through the forms of these cultural artefacts. Power, then, is harnessed to reference both 'literal' power – in the sense that we might say 'power' meaning energy – and more nebulous channels of power, for example, discursive, legislative, subversive power. This multi-valency of power is, like the concrete material used to harness the waters behind the dams, both localized and globalized.

Whilst electricity was previously understood in terms of a public good or service, Kate Showers notes that it 'was transformed by neoliberal economics into a commodity, or private asset, governed by market forces for private profit' (Showers 2011: 1669).[5] For Showers's case study – the Grand Inga project on the Congo River – this means that the electricity generated could be transported, via a network of HDVC (high-voltage direct current) lines, to faraway places, notably those with the requisite financial means, for example, in Europe and the Middle East. Grand Inga was thus a project of development, particularly for privatized South African electricity utility Eskom Enterprises, to 'lift its people out of poverty and deliver sustainable development' (Eskom Enterprises's Chairman Reuel Khoza, as quoted in Showers 2011: 1669) by displacing electricity as commodity. Patrick Murphy makes a similar argument in his essay 'Damning Damming Modernity'. 'Megadam construction', he argues 'is based on the notion of taking the electricity and the water to urban, high consumption cities and large industrial projects, such as mining operations, at the expense of indigenous communities who historically migrated to the water' (Murphy 2011: 31). These communities are then forced to migrate again to the water, now in urban centres, 'where it [the water] and the power produced by it are salable commodities' (ibid.). Power organizes communities. As projects such as these qualify as 'green' projects, they can trade carbon credits under Kyoto protocol, thus magnifying the shifts of power (by, for instance, externalizing costs and risks of consumption from some parts of the planet to others).

---

[5] As Mimi Sheller compellingly argues, Iceland exports its cheap electricity by means of the condensed form of aluminium (cf. Sheller 2014: 205–20). Here we have an understanding of electricity as tradable commodity or asset, rather than as service.

Thinking dams means rethinking energy. Whilst the water of hydroelectricity is a renewable (if unpredictable[6]) resource, this does not make the dam a source of 'green' energy. Megadams are themselves in fact a source of greenhouse gasses, specifically carbon dioxide and methane. Some studies have indicated that reservoir emissions are not only higher than assumed, but that in the tropics they can even exceed emissions from equivalent fossil-fuel power plants (cf. also McCully 2001: xxxii–xxxvii). This is primarily a function of rotting matter, condensed in the reservoirs: Whilst the assumption is that the emissions will reduce after the initial biomass displaced by the inundation of water behind the concrete dam has decomposed, this does not account for the emissions released with the discharge of waters from the turbines and spillways (cf. ibid.: xxxiii), nor does it provide for the inundation of methane and carbon dioxide sinks by the waters of lakes produced by the installation of dams. And, further, the production of the materials of the dams themselves – the concrete, the steel – also produces emissions. A tonne of cement, a key ingredient in concrete, generates almost a tonne of carbon dioxide in production (cf. Rubenstein 2012).

Like the misunderstanding that leads consumers to become annoyed at organic food that has been wrapped in surplus plastic or shipped in across the globe (a conflation that comes from the quality 'organic' not being interchangeable with 'environmentally friendly'), renewable energy is not the same thing as environmentally sound energy. Nor is it the same thing as socially sound energy.

At this conjuncture, a World Bank report called 'Progress toward Sustainable Energy 2015' warrants brief analysis for the mechanisms through which its findings are presented. The document comprises mostly lengthy written reports and statistics rendered in an array of different graphs. In addition to one photograph of the Secretary General included in his keyword, there is a set of photographs in the key findings section I want to examine in more detail: three of the four photos in this section show humans in quotidian scenes.

1. A Black woman in African dress is shown in front of a store, with the caption 'Energy has a key enabling role in food security and nutrition' (World Bank 2015: xi);
2. a young (PoC, person of colour) child is shown having their teeth examined by a male (PoC) dentist (with an assistant, probably male,

---

[6] Particularly given large-scale shifts in rain and snowfall patterns as a consequence of (anthropogenic) climate change, water flows are becoming increasingly unpredictable. As McCully succinctly notes: 'Climate change is rendering obsolete one of the key assumptions used in dam planning and design – that the hydrological past is a reliable guide to the hydrological future' (McCully 2001: xxxvii).

looking on in the background) captioned with 'Modern energy provision is a critical enabler or universal health coverage' (ibid.: xiii);
3. two (PoC) women wearing saris are shown seated at sewing machines, with a further woman, possibly a supervisor, standing nearby; the image is captioned 'Access to affordable energy services can reduce both time and effort spent in productive labor' (ibid.: xv) and
4. in the only photograph that does not depict humans in this section, a hydroelectric dam in aerial shot, captioned with 'Energy and water resources are inextricably tied together' (ibid.: xiv).

The captions together with the photographs offer faces, or interfaces, for the visualization of key goals of energy production. Development goals – access (to services, to health and to the workforce), affordability and stability – are given a visual cue, faces stressing their importance. And, crucially for this chapter, dams are afforded a status alongside them.

In an earlier report from 2000, the World Commission on Dams provides as the first two of its four central findings, that 'dams have made an important and significant contribution to human development, and the benefits derived from them have been considerable' (WCD 2000: xxviii) and, at the same time, that 'in too many cases an unacceptable and often unnecessary price has been paid to secure those benefits, especially in social and environmental terms, by people displaced, by communities downstream, by taxpayers and by the natural environment' (ibid.). The amount of funding channelled and stored in megadams suggests that it is a powerful index of development – both a sign-of and a sign-for material practices of engaging with the resources of the globe, engendering responses both positive and negative. Kate Showers, for instance, has also argued (from an African context): 'For both colonial regimes and independent governments, large dams were symbols of a government's strength as well as essential infrastructure. As symbols they were often contested politically' (Showers 2011: 1657). The dam is thus figured as an index of development. Juxtaposed with the images of people, it is presented as a 'face' of sustainable energy in the 2015 World Bank report.

If Jawaharlal Nehru has suggested that dams are the temples of modern India,[7] and if the international community rallied monetary support in order to shift the temples of Abu Simbel – more support was garnered for the temples than for the approximate 100,000 peoples displaced by this same construction – there is

---

[7] Nehru later retracted this bold statement; however, its discursive power continues.

clearly a dynamic of sacred place, as well as an emplacement of development, at work in dam discourses. In a move that echoes these concerns, Edward Abbey brought comparisons of the Glen Canyon walls together with the cathedrals of Europe (Brinkley 2006: xix). It is to the various representations of Glen Canyon Dam to which I now turn as a central case study for this chapter.

Explosions taken for fireworks: Glen Canyon Dam

The Glen Canyon Dam I analyse here is one that emerges out of a 'layering' of texts: Edward Abbey's *The Monkey Wrench Gang* and Leslie Marmon Silko's *Almanac of the Dead* – two widely received texts – are joined by documentaries by Earth First! and the United States Department of the Interior's Bureau of Land Reclamation, folk songs and a thriller (Gary Hansen's *Wet Desert*), as well as a range of popular movies and newspaper articles. This 'layering of texts' comprises a minor catalogue of stories-so-far (cf. Massey 2006). The order, in what follows, is not teleological in the sense of a trajectory moving from simple to complex. Rather, the stories map onto and across each other, forming a dense coagulation of meaning, analogous to the concrete materials of the megadam itself. Through these texts, different sentiments and sediments emerge: Although they draw on each other, they also anticipate each other, in strange ways.

Accordingly, and to stress such an interpretation, I refrain from analysing them in specific chronological order. Whilst literary and cultural texts *do* reference each other, and the examples I draw on also do this, to insist on a linear chronology elicits, on the one hand, a comprehension for the 'first' text as somehow original and originary, suggesting somehow that consequent representations are necessarily derivative. On the other hand (and somehow-but-not-quite contradicting this), such a linear chronology evokes an understanding that the consequent representations are somehow more complex, as they draw on previous representations. Such a linear understanding of intertextual reference would place the interpretative authority with the author and ultimately leads to interpretations that are limited to generating catalogues of references. In what follows, I therefore stipulate an intra-textual interpretation by refusing such chronology in order to engender a comprehension of the Glen Canyon Dam that emerges through a 'layering', or, to employ a more fitting and fluid metaphor, a 'coalescing' of representations. This is, in Karen Barad's terms, a reading for diffraction rather than reflection (Barad 2007: 89), or, to use the metaphor of flowing water, a reading for confluence rather than influence. In the sense that a setting might evoke not only a background but also a settling – a form through which materials take shape and take place – the texts that assemble around the Glen Canyon Dam also assemble the Glen Canyon Dam.

To the site: The Glen Canyon Dam is a structure typified as concrete thick arch-gravity dam. It is an infrastructural node, one of a series of dams on the Colorado River,[8] and is situated in Arizona near the border to Nevada in the west of the United States. The dam was constructed in order to harness water, following the Colorado River Compact of 1922. This decree divided up water as a resource amongst six US states and, importantly, the nation state of Mexico, through which the Colorado River passes before it reaches the Pacific in the Gulf of California.[9] Construction began in 1956 and was not completed until ten years later. The reservoir created by the construction of the Glen Canyon Dam was named 'Lake Powell' for the explorer, anthropologist and geologist John Wesley Powell, who was the first director of the Bureau of Ethnology at the Smithsonian Institute.[10] The affected region has Anasazi sites and also was home to Navajo and Paiute peoples.

The waters of the Colorado river harnessed by the Glen Canyon Dam are also used in 'non-renewable' energy production: The Navajo Generating Station, located approximately 10 kilometres away from the dam, is a coal-fired electrical generation plant that uses the water of the Glen Canyon Dam to cool its systems. Whilst the station provides power to a number of southwestern cities, the bulk of power produced by the plant is harnessed for distributing the water of the Colorado River dams throughout the region. The Navajo Generating Station is, then, a coal plant constructed to circulate water from a river used for the production of hydroelectricity and which was given the name of the Native American tribe displaced by the construction of the dam. The forced removal and decimation of First Nations or Native Americans through the various phases of North American colonization and burgeoning nationhood is symptomatic of megadam construction.

---

[8] Including the Hoover Dam, depicted in Ansel Adams's photographs, located close to Las Vegas in Nevada.

[9] As Jonathan Waterman notes in an op-ed piece for the *New York Times*, 'Where the Colorado Runs Dry', 14 February 2012, and in his earlier book *Running Dry* (2010), the river itself does not always flow into the Pacific, for the water it holds is diverted according to the Colorado River Compact, which only afforded Mexico a portion of the water at a later date.

[10] Powell's racist views were informed by a classification of human societies into 'savage', 'barbaric' and 'civilised', according to their use of technologies (cf. Haller 1995: 108). The adoption of his name for the lake is 'telling' in its evocation and perpetuation of technological development as the *sine qua non* of megadam construction.

I begin my Glen Canyon Dam analysis somewhat obliquely, with Jared Farmer's *Glen Canyon Dammed*. Farmer's account, published in 1999, comes chronologically *after* most of the texts I address later. This cultural history of the dam attempts to offer a 'well-rounded' account of the site, by eliciting the opinions and stakes not only of those who resist the dam, but also of those who support its continuing existence. Where most of the cultural texts that take the Glen Canyon Dam as site resist the materiality and symbolism of the dam (in activist and environmentalist modes), Farmer's approach mobilizes other stakeholders, such as farmers and recreational users.

In a chapter entitled 'Drain This?', for instance, some points, such as the concentrations of heavy metals and toxicants in the accumulated sediments of Lake Powell, leaking from motorboats and leaching from farms and uranium tailings pits, are mobilized as arguments for and against the Glen Canyon Dam. Binary options emerge: get rid of the lake, stop the problem now versus leave the lake and keep the problem at bay (Farmer 1999: 184–5). Farmer also notes that the dam's waters are instrumental in the cooling of the Navajo Power Station (see above), which employs some 1,200 Navajo tribal members (cf. ibid.: 185). Farmer is also explicit in referencing houseboat recreation on the reservoir (which also comprises one of the narrative strands in Gary Hansen's thriller, *Wet Desert: Tracking Down a Terrorist on the Colorado River*),[11] suggesting that such engagements with the Glen Canyon Dam reach the order of millions of people who visit the lake. Against the more vocal opponents of the dam, which comprise the bulk of responses, Farmer mobilizes stakeholders in such recreational uses, suggesting that whilst they might not 'fathom … the depth of this "spiritual" feeling"' (Farmer 1999: 189), that this nevertheless 'cuts both ways': 'anti-dam activists generally fail to appreciate the importance of Lake Powell to others. Because the sin of the dam is to them self-evident, they do not accept the love that others profess for the reservoir. They do not respect the recreation that goes on there' (ibid.).[12] Farmer, in an attempt to articulate a well-rounded account by giving voice to resistance from stakeholders on both sides, finds difficulty in broaching the division. His vocabulary, for instance 'anti-dam activist', makes

---

[11] Hansen's novel establishes an engineer, Grant Stevens, as its hero (who is unhappy that the heyday of megadam construction has passed). As the subtitle suggests, the activist is figured as a terrorist: The novel's sympathies are clearly delineated. A series of explosions reverberate through the novel, detonations destroying the dam walls threaten the livelihoods of downstream farmers, as well as the lives of the houseboat dwellers. Stevens, once engineer, now detective, ultimately finds the perpetrator, who is able to escape once more. Our 'hero' is rewarded for his diligence by being tasked with reconstructing the dam.

[12] A passage from *The Monkey Wrench Gang* jostles similarly with the claims of stakeholders:

such differences in opinion more concrete, setting up resistance rather than clearing grounds for settling matters differently.

Shifting media, a different perspective: In the short documentary film entitled 'The Cracking of Glen Canyon Damn with Edward Abbey and Earth First!' from 1982, Edward Abbey entertains the tropes of alienation and invasion to describe the 'fantastic structure' of the Glen Canyon Dam, shown in the background of the opening scene ('The Cracking').[13] Abbey, twirling his fingers through a barely visible thread and gesturing to the Glen Canyon Dam in the background, declares: 'I see this as an invasion. This looks like creatures from Mars to me. I feel no kinship with that fantastic structure over there'. Abbey is insistent, repeating: 'No sympathy with it whatsoever' ('The Cracking': 00:40–00:58). This video, an excerpt from a longer piece entitled 'The Cracking of Glen Canyon Damn', is exemplary for the interactions between aesthetics and politics, as a site and node of entangled development (as the thread itself might suggest), and accumulations of meanings through various stories.

Abbey is shown giving a speech from the back of a utility vehicle when cheers can be heard as the protestors unfold a long strip of black plastic over the side of the dam: a 'crack' on the face of the dam. Another shot, again taken at a distance from the air, interposes the cheers of a handful of protestors with the visuals of the unfolded 'crack' and the folksy music of Johnny Sagebrush (Bart Koehler) singing 'Were you there when they built Glen Canyon Dam? / Were you there when they killed this river dead?'[14] As the aerial shots gain distance to the site of the Glen Canyon Dam, a suburban development enters the scene. The small plots of bright green – trees and gardens, even larger plots of lawn – intersperse the dark grey lines of road. This shot depicts, and construes, the flow of water, the flow of power and the flow of development.

Using wide shots taken from the air, the film portrays the vastness not only of the dam, but also of the riverine landscapes created by the dam. The contrast

> The doctor was thinking: All this fantastic effort – giant machines, road networks, strip mines, conveyor belt, pipelines, slurry lines, loading towers, railway and electric train, hundred-million-dollar coal-burning power plant; ten thousand miles of high-tension towers and high-voltage power lines; the devastation of the landscape, the destruction of Indian homes and Indian grazing lands, Indian shrines and Indian burial grounds; the poisoning of the last big clean-air reservoir in the forty-eight contiguous United States, the exhaustion of precious water supplies – all that ball-breaking labor and all that back breaking expense and all that heartbreaking insult to land and sky and human heart, for what? All that for what? Why, to light the lamps of Phoenix suburbs not yet built, to run the air conditioners of San Diego and Los Angeles, to illuminate shopping-center parking lots at two in the morning. (Abbey 2006: 173)

[13] The short-version video 'The Cracking of Glen Canyon Damn with Edward Abbey and Earth First!' is available online. It is suggested that this is one of Abbey's rare live interviews.
[14] Cf. also Barber (1996): 135; for an analysis of the rhetorical figures used, see Ross (2015).

between the yellows and browns of the land with the deep blue of the river and lakes is stark: Where once there were rifts in the environment, deep canyons forging their way through an arid landscape, now water and land are level, stretched into a smooth plane behind the dam. A horizontal – and vast – vista. The documentary excerpt closes with the skeletal power lines against rosy clouds, invoking aesthetics similar to those of Ansel Adams. The film, by EarthFirst!, mobilizes the myth of Edward Abbey's 'monkeywrenching' from his popular novel *The Monkey Wrench Gang*, and the celebrity of Abbey as author, in its video (cf. Barber 1996: 134–6).[15]

The explosions mistaken for fireworks of the title of this section references a key scene in the prologue of Abbey's novel. The crowds assembled around the Glen Canyon are waiting ('The people wait. ... Waiting, waiting. ... The Indians also watch and wait. ... The citizens wait' (Abbey 2006: 2–3)).[16] The cutting of a ribbon finally takes place and a 'flurry of sparks which followed as the ribbon burned' send a 'rash of fireworks' blazing up the walkways: 'Whirling dervishes of smoke and fire took off and flew, strings of firecrackers leaped through the air like smoking whips, snapping and popping, lashing at the governors' heels. The crowd cheered, thinking this the high point of the ceremonies. / But it was not' (Abbey 2006: 5).

But it is not fireworks that cause the centre of the bridge to rise up, to break in two, they do not cause the 'monolithic sandstone of the canyon walls' (ibid.) to shake, or bring 'fragments and sections ... to fold, sag, sink and fall, relaxing into the abyss' (ibid.: 6). Instead of effervescent flashes of light, objects are launched into space by a bomb, amongst them, 'gilded scissors, a monkey wrench, a couple of empty Cadillacs' (ibid.). And thus a monkey wrench is launched, an explosion (mis)taken for fireworks.

Abbey, in *The Monkey Wrench Gang*, has his 'ecoterrorists' destroy the Glen Canyon Bridge (and not, actually, the dam). Also a concrete infrastructure, the bridge 'represent[s] corporate and governmental disregard for the land on which we live' to the monkey wrench gang, but, 'a bridge also represents a freedom of access, the possibility of connections, union' (Barber 1996: 134) as Katrine

---

[15] Abbey's novel is also invoked in Earth First! Newsletters and fundraising materials. A further link between Abbey's novel and the environmental group Earth First! is also evident in the title of David Foreman's book *Ecodefense: A Field Guide to Monkeywrenching*.

[16] Here, the parsing of the sentences suggests that the 'Indians' are neither 'the people' nor 'the citizens'. Whilst an argument could be made that they might constitute a specific audience, and are thus foregrounded, this only has the effect of normalizing an implicit whiteness of the remainder of the audience. A troubling nationalistic sentiment of the United States permeates the text, even as it propounds its various countercultural/activist stances. The voices of Abbey's novel, the narrator and characters alike, have misanthropic, racist and misogynist tones.

Barber argues in her analysis. The destruction of the bridge, following Barber, is a divisive action as much as an act of 'ecotage'. At the 'end of the day', the Glen Canyon megadam itself is left standing in the imagined world of Abbey's novel.[17]

The explosion taken for fireworks of this section's title is also a story of (mis)taken identity: of meanings attributed to signs that belie, and reveal as lie, intentions.[18] It picks up on the rhythms and impetus of Homi Bhabha's 'Signs Taken for Modernity', thus alluding to issues of modernities and development.[19] In other renditions of similar events, the symbolism of the explosion – the act of incursion into the monolithic renderings of the state's power – shifts from bridge to dam, from infrastructure of connection to infrastructure of power.

Explosions taken for fireworks, then, are (deliberately) mis-read attributions of intent, taken up in a myriad of other texts that assemble and assemble around the Glen Canyon Dam. Another such text is Leslie Marmon Silko's *Almanac of the Dead*, published in 1991. Leslie Marmon Silko identifies as a Kawaik (Laguna Pueblo) writer. In one of her readings of Silko's *Almanac of the Dead*, Joni Adamson notes that 'indigenous attempts to bring earth-beings into politics in opposition to massive hydroelectric projects or vast logging or mining operations often get labelled "ethnic politics", a phrase used to dismiss the legitimacy of indigenous understandings of a more plural world' (Adamson 2014: 187). I will return to this point later. First, I want to examine closely the figuration of Glen Canyon Dam in this novel. The scene of Glen Canyon Dam collapse in *Almanac of the Dead* is mediated through an ekphrastic video, described within the closing pages of the novel. A group called Green Vengeance, described as

---

[17] 'The windmill Abbey wanted to tear down the most' Douglas Brinkley argues in his introduction to *The Monkey Wrench Gang* was the 'concrete colossus' that 'had stemmed the natural flow of the Colorado River, desecrating the steep walls of the magnificent Glen Canyon that Abbey imagined grander than all the cathedrals in Europe' (Brinkley 2006: xix).

[18] Douglas Brinkley places Abbey's novel in a pattern of folk influences by suggesting that one of his inspirations for heading West was listening to Woody Guthrie (cf. Brinkley 2006: xvii). Woody Guthrie was a singer-songwriter and folk musician, known for his social justice songs, like 'This Land is Your Land'. And yet Woody Guthrie was employed as a research assistant to 'write some songs in praise of the dams' (Reisner, as quoted in Lindholdt 2015: 121).

[19] Homi K. Bhabha's essay 'Signs Taken for Wonders: Questions of Ambivalence and Authority under a Tree outside Delhi, May 1817' alludes to the effects of the translation and transposition of the 'English book' (also called the 'Holy book', i.e. the Christian Bible). What happens to a text once it passes across cultures? Or, indeed, when it is repeated? As Susan Bassnett and Harish Trivedi point out, most modern Indian languages (Bhabha's 'native' turf) use the word *anuvad* from the Sanskrit for the English term 'translation': 'The underlying metaphor in the word *anuvad* is temporal – to say after, to repeat – rather than spatial as in the English/Latin word translation – to carry across' (2002: 9). The double allusions of translation – once temporal, once spatial – are helpful for considering the function of modernity – once temporal, once spatial. Bhabha's 'English book' proves, at the end of the essay, to be an object as much as a signifier, evinced in the rage of a missionary who notes its use as waste paper or as an object for bartering (cf. Bhabha 1985: 163–4). For a further discussion of material modernities, cf. 'Introduction'.

'eco-warriors' (i.e. Silko 1992: 726), take the stage at the International Holistic Healers Convention to present a video of their action.[20]

Showing 'a jerky sequence, videotaped from a moving vehicle', the video has been stripped of sound to concentrate on the 'show' of destruction. Landscapes of 'brilliant burnt reds and oranges of ... sandstone formations' are juxtaposed against 'dark green juniper bushes' before cutting to 'interiors of motel rooms with figures in ski masks and camouflage clothing standing by motel beds stacked with assault rifles and clips of ammunition' (Silko 1992: 727). Shots are described as being framed carefully so as to avoid showing faces: 'Next, the screen had been filled with highway signs and U.S. park service signs; in the background was the huge concrete mass that had trapped the Colorado River and had created Lake Powell. / GLEN CANYON DAM; the sign had filled the entire screen' (ibid.). The shot moves out again, allowing the 'concrete bulwark of the dam' (ibid.) to come into focus, in front of which small figures dangle precariously, unbeknownst to any of the depicted by-standing tourists or employees. Zeta, the focalizer of this passage, likens the image to that of spiders when the 'giant video screen itself appeared to crack and shatter in slow motion' covering the figures in dust and smoke, before the 'entire top half of the dam structure ... folded over collapsing behind a giant wall of reddish water' (ibid.).

Silko's 'eco-warriors' tell the crowd the official version: 'A massive structural failure due to fault asymmetries and earth tremors' (ibid.), but the crowd at the International Holistic Healers Convention roar and cheer again as the person on stage draws on the rhetoric of martyrdom to describe the deaths of his fellow warriors. Another companion, whose tone is calmer, frames such actions as acts of war, against the state, and for the earth. The medialized performance of the explosion of Glen Canyon Dam in *Almanac of the Dead* thus accomplishes what neither Abbey's *The Monkey Wrench Gang* or the Earth First! videos manage: The concrete dam is no longer.[21]

What, then, to make of Silko's account, albeit brief, of the explosion of Glen Canyon Dam? The deliberate staging of destruction recollects the video of Earth

---

[20] The denomination as 'eco-warriors' here, like the phrase 'eco-terrorists' elsewhere, speaks volumes to the discursive renditions of statehood and state-sanctioned versus state-interdicted violence. Through such terminology, resistance is rendered illegal (but not all aggressions). Linda Hogan's *Solar Storm*, as discussed in my introduction, is another novel that depicts North American indigenous resistance towards dams. There, a similar point of the unevenness of access to (political) power is articulated when the 'activists' note that 'reversing the truth, they [the powers that be] would call us terrorists' (Hogan 1997: 283) and 'if we'd done such a thing to them, we'd be arrested and hauled away' (ibid.: 287).

[21] The scene in *Almanac of the Dead* resonates with Arundhati Roy's suggestion that 'dams are not newsworthy until the devastation they wreak makes good television. (And by then, it's too late)' (Roy 2006: 307). I come back to this intersection of spectacle, slow violence and Roy's essays later;

First!, accomplishing on a deictic level what the Earth First!-ers can only allude to with the sheets of black plastic in their video. Obviously, the timeline suggests that the Earth First! video might be an important precursor, having been filmed some ten years prior to Silko's novel's publication (and the video, then rather obviously, being a completion of Abbey's novel, which stops short of exploding the Glen Canyon Dam). However, such an interpretation closes down paths of influence, critiqued earlier for its implicit teleology and emphasis on origins, and, at the same time, it draws on speculation; this it is certainly not the most pertinent critical move to be made at this juncture.

Rather, at the same time as evoking and entangling numerous discourses of power, *Almanac of the Dead*, together with the other texts, reveals a predilection for the particular site of the Glen Canyon Dam: Each time the Glen Canyon Dam is revisited, its story is not *only* continued, but thickened, layered, condensed. The fascination is *not* simply a function of the precursors, nor is it writers/artists 'picking up leftovers' from previous cultural artefacts. These are *imaginings* and not histories in a true sense, but these imaginings *do* contribute to the histories in a very real sense. The history of Glen Canyon Dam is rendered multiple, not as a culmination of a singular historical trajectory, but instead a rendering of the various stories of Glen Canyon Dam.

Another aspect that makes the explosion of the Glen Canyon Dam in Silko's *Almanac of the Dead* pertinent to this chapter is its evocation of power dynamics. Amongst the viewers of the screening at the Convention is activist-hacker Awa Gee, who 'believed very soon these last remaining eco-warriors would push forward with their plot to turn off their lights. ... Awa Gee decided he would help the eco-warriors turn out the lights, although they might never even know Awa Gee's contribution' (Silko 1992: 730). A digital 'protovirus' Awa Gee had already developed would help him accentuate any fluctuations in power, the readers are told, and he would be 'counting on the "cost cutting" of the giant power companies to curtail or cancel auxiliary emergency systems. But if the plan worked, if the lights went out all over at once, then the United States would never be the same again' (ibid.). Power, here, is rendered an entangled mess of electrical and political flows.

This particular thread of *Almanac of the Dead* – jostling amongst several other plots that follow drug money, cartels, other figurations of counter-state incursions – is joined by plot-lines that trace individual and collective

---

it also resonates with the definition of infrastructures becoming 'visible' only upon destruction (see above with references to Rubenstein, Robbins and Beal 2015).

disenfranchisement.[22] Silko's novel has thus been characterized as 'overtly and often uncomfortably political' (Tillett 2014: 5), with 'its insistence that readers repeatedly unsettle and relocate themselves, culturally, geographically, even temporally' (ibid.: 6). Tillett's brief survey of responses to the novel elicits responses united in the forcefulness of their responses, thus picking up on the themes of power, empowerment and displacement stressed throughout this chapter. If Tillett suggests that *Almanac* might be situated within 'a history of multiethnic American literary resistance and revisionism' (ibid.: 9), then this might productively resonate with the idea of a resistor in physics, or as forces that hold up (against) power, in the sense that the dam, which is stressed in my interpretation, also restrains and releases power.

The Glen Canyon Dam thus comprises a site of negotiations of power and resistance and of their material manifestations. The multiplicity of representations offers a site for rethinking through the overlap between 'material ecocriticism', 'material cultures' and 'new materialism', as discussed at length in the introduction to this book. These dimensions extend into the textual corpus that comprises the imaginings of Glen Canyon Dam (and, as will be shown in the next section, other sites): the materiality of beginnings and endings is crucial in such texts, both those that suggest its destruction and also those that celebrate its construction.

The United States Department of the Interior Bureau of Reclamation video 'Operation Glen Canyon' (1961) commemorates the construction of the Glen Canyon Dam. Like the Earth First! video analysed at the outset of this section, it avails itself of the use of aerial shots. Unlike the Earth First! video, the tone of this governmental video is decidedly celebratory, portraying the grandeur of engineering power.

In the opening sequence, before the title of the 28-minute film is superimposed, the sound of a propeller aeroplane is audible. The rotation of the propellers is echoed later by the rumble of diesel engines powering earth-moving vehicles. The auditory channel thus forge links between various feats of (mechanical) engineering, that is, between plane and dam. The viewer follows the plane, as it follows the plane of the land, dipping into the canyon. The voice-over begins by placing the Colorado River in a watershed, comprising the 'towering mountains

---

[22] Laura Shackleford brings together the various strains of the narrative within the context of subaltern resistance to capitalism: 'Joining forces with an army of the homeless, with Vietnam veterans, and with ecoterrorists, the Native American and other indigenous characters in the novel comprise key nodes in what is explicitly envisioned as a broader subaltern network of resistance to the structural inequities of global capitalism' (Shackleford 2006: §38).

of Colorado, ... the rugged mountains of Wyoming, ... the lofty mountains of Utah, ... the rolling mountains of New Mexico' (USBR 1961: 1:51–2:04). US American states are evoked in this description, and later (cf. ibid.: 2:52–2:54), providing a context of state-based livelihoods, for the images of deserts and forests presented in quick succession. This reinforces the idea of emplacement of the Glen Canyon Dam in a particular site. However, several moments in the documentary show the limited extent to which this spatiality is considered: for instance, outside of the opening sequences there is no further mention of Mexico, through which the Colorado flows into the sea.

The initial descriptions, visual and oral, are of nature. Only after this framework has been extensively invoked – by means of a series of shots that extends for the first three minutes – are humans introduced into the landscape: First, a shot that depicts men looking down at the canyon, later posed crouched over maps and plans. These men,[23] the voice-over announces, are 'rewarded for their efforts as the first blast began the construction of the giant Glen Canyon Dam' (ibid.: 3:30–3:38). A tooting sound, an acoustic mix of the reedy oboe (picking up on the symphonic soundtrack of the nature images) and the blast horn, corroborates the voice-over as the viewer witnesses a blast on the side of the dam. Nature is figured as resource, as raw materials for harnessing – and displaying and displacing – power. A cloud of dirt intrudes into the canyon, the echoes of the blast are presented without any commentary or background music: front stage, then, for development.

And thus, given the oblique chronology of this analysis, the blast echoes back through those other blasts that would see the destruction, rather than construction, of the dam.

The construction workers, an 'army' (ibid: 4:04., also 24:30), emerge into the 'land of the Navajo' (ibid.: 4:13–4:15), themselves a trace presence on the screen. The first non-white humans visually represented in the film accompany the spectacle of the first pour, suggestive of Rob Nixon's unimagined communities (see below). They are backgrounded to the materials of construction: This is a film concerned with the spectacle of engineering. Diggers, caterpillars and an array of large construction vehicles and then a further blast as the canyon

---

[23] The language is gendered in this way in the film, perhaps a function of the early 1960s when it was made, but also echoed in the visual representations of humans, which are, as far as they can be rendered in terms of gender, masculine in appearance, dress and movements. Notably, the first woman in the documentary is at a kitchen table, feeding children (USBR 1961: 12:13–12:23), that is, firmly established in a domestic scene. I address gendered divisions into public/private space, and the dualistic repercussions of such divisions, in conjunction with the 'Saran Wrap' advertisement in the next chapter (Chapter 4).

walls collapse into the crevice. Earth is moved, removed, the narrator tells us, in actions that 'remove the rock, remove the sand, remove the gravel laid down in the ages before man. Get down to the bedrock and wash it clean' (ibid.: 11:12–11:20) so that it is 'ready to receive the massive concrete of Glen Canyon Dam' (ibid.: 11:32–11:36).

In the film, the earth is clearly figured as receptacle for the improvements of 'man', that is, as resource. The concrete, as mixture of the prevailing elements of earth and water that figure so prominently in the visual aesthetics of this 'Educational Documentary', together with the 'improvements', that is, the steel of the bridge and metals of the earth-moving vehicles, provide a stage for development, explicitly evoked: 'As actors in the drama of the Colorado, the workmen in the canyon below now hold a stage' (USBR 1961: 25:18–25:24). The spectacle of the 17 June 1960 for the 'first bucket of concrete' (ibid.: 17:38–17:41) is accompanied by a triumphant big band performance as we are presented with the journey of cement, frozen water and aggregate along long conveyer belts to the site of the dam. The first bucket is set in scene as a tightrope artist, precariously dangling from a set of pulleys high above the ground of the canyon. Again, we are shown a spectacle of engineering, evoked through repeated images of materials in motion: being poured, travelling atop conveyer belts, churning, falling.

Materials in motion, sourced locally and further afield, are emplaced, constructed, to form the dam. We learn: a large plant, six miles from the dam site, washes and screens the stones and sand that will comprise the aggregate of the concrete. The aggregate, transported on the backs of trucks, is released into underground chambers for storage, before being transported to the site of the dam via underground conveyer belt. The image is interesting: trucks returning dirt to dirt, earth to earth, as the bellies of the trailers are opened to release the sorted aggregate into the earth. Cycles of material are evoked through water, which, the documentary informs us, also requires a plant (ibid.: 16:06). This is a refrigeration plant, which freezes some of the water to be used in the concrete mixture. It also produces cooled water to run through the concrete of the dam to carry away some of the heat as the concrete of the dam cures.[24] As the aerial shots of the opening scene are echoed through a pan out, no longer accompanied by the sound of the propeller, the film closes with the observation that 'Glen

---

[24] Other materials are sourced further away. The cement is produced at a plant 188 miles away; the pozzolan, which produces 'desirable qualities' to supplement the cement, is obtained from a plant near Flagstaff (Arizona, USA).

Canyon Dam will endure, an important and noble venture for tomorrow' (ibid.: 27:55–28:00).

*　*　*

The metaphoric qualities of the dam, as index of development and as site of resistance to the power the dam generates, shift in scale through the various texts that elicit the site. The texts, together, evoke a node of infrastructure whose significance extends its materiality; however, it is through this materiality that this significance is portrayed. Glen Canyon Dam figures texts as entanglements, as a simultaneity of stories-so-far, through its concreteness and its concrete. By *not* placing the texts in chronological order, I work to evoke a sensitivity for the multiplicities and entanglements of the various texts that coalesce around Glen Canyon Dam.

As a formable material, concrete constitutes a reminder of the co-constitution of environment through various agencies: the agencies of the designer, engineer or architect, of funding bodies, of workers on site; of users and abusers; and of the various responses to environments encoded through various texts (policy documents, novels, visual artwork (graffitis, photographs), and audiovisual films (in particular, documentaries)). The local and global (or trans-regional) come together through the materials of Glen Canyon Dam. Concrete not only assembles entanglements of construction (and destruction), then, but also of critique.

The energy released by the explosions – making way for the construction of the dam, the (imagined) destruction of the dam – brings to the fore the material incursions of the dam. Power is released, contained, construed with the concrete walls of the megadam. Again, by rearranging the chronology of the texts, specifically, by not beginning with the United States Department of the Interior Bureau of Reclamation (USBR) video 'Operation Glen Canyon', I resist the teleology of causality and effect running back through the texts to the representation of its material construction. Such an organization would afford the video an originary function, suggesting that the other texts *inevitably* respond to this one and close down the interpretation of its materiality to such a causality. My reading stresses relationalities rather than mimesis, entanglements rather than separation, and derives from an understanding of the inherent physicality of the world that does not insist on subjects and objects, but rather their co-constitution. The multiple explosions of the texts of Glen Canyon Dam comprise a spectacular reverberation: materialities in action.

Place troubles: Local and global 'Indians'

The mediality of materialities, crucial to the entanglements of the site of the Glen Canyon Dam as the above analysis indicates, suggests a dynamic of power, with specific entanglements and interruptions. These may materialize as explosions, as above, that is, as spectacles. Rob Nixon, in his influential study *Slow Violence*, includes a chapter entitled 'Unimagined Communities: Megadams, Monumental Modernity, and Developmental Refugees' as part of his project of articulating a sensitivity towards dominant modes of visuality in the mediation of dynamics of power. He stresses the workings of slow violence as 'neither spectacular nor instantaneous, but rather incremental and accretive' that necessitates critical approaches that 'engage the representational, narrative, and strategic challenges posed by the relative invisibility of slow violence' (Nixon 2011: 2), and which is 'often not just attritional but also exponential' (ibid.: 3, cf. also Chapter 5). Further, the people who are 'the principal casualties of slow violence' (ibid.: 4) are those whose 'unseen poverty is compounded' (ibid.) by slow violence's invisibility. In addition to the evocations of the spectacular blasts of megadam construction and (imagined) destruction of Glen Canyon Dam above, it is important to stress the slower, attritional layerings of effects (rather than the layerings of representations). In what follows, I address the indexical quality of dams as development to investigate the capacity of agencies of resistance to assemble by way of the counter-intuitive conflation of 'inconvenient Indians'. Issues of place and displacement are crucial in this regard.

Megadams, again, are part of a larger project to harness resources, in particular water-as-electricity but also water-as-water, that is, as resource for agricultural projects. This harnessing of resources is enacted in the name of energy and generates not only power but also displacement, of resources and of livelihoods. 'Livelihoods' is evoked here to suggest not only social, cultural and economic patterns of human existence, but also the myriad ways in which these factors intra-act with the material world that co-constitutes them (Barad 2007, cf. also 'Introduction'). Power is generated (and, as shown earlier and later, resisted) through such incursions and intra-actions and is patterned through texts that constitute and resist this generation of power.

Just as power has a complex relation to disempowerment, place is also troubled by displacement. Livelihoods are displaced by the lakes and reservoirs formed by the dams, and with them, patterns of inhabiting and comprehending the environments are also altered. According to the World Commission on Dams,

40–80 million people have been displaced (cf. WCD 2000: 16), and 60 per cent of the world's rivers have been affected, by dams and diversions (cf. ibid.: xxx). Through these figures – abstract as they may seem in a policy document – stories of resistance and development unfold.

The construction of megadams, as well as smaller dams, entails shifts in the distribution of water and results in the production of large lakes. Concrete structures structure space: like other infrastructural projects, dams can be considered terraforming constructions. Peoples were and are forcibly displaced, often concomitantly with expansion of empires, from and also towards regions deemed attractive for particular agricultural, mining and other extractive or profit-promising projects. Similar patterns of exclusion are evident in the land-clearing gestures that are concomitant with the construction of megadam projects: Max Liboiron, drawing on the work of la paperson, outlines at length how the externalization of land creates sinks – spaces that are abandoned to pollution – as a crucial component of the colonial project (cf. Liboiron 2021: 81–111). Like the materials that are displaced, the specific assemblage of livelihoods in a potential future *after* the megadam is both speculative and of a duration that exceeds the sudden incursion of the concrete (I will go into more detail on the shifts of timescales with materials in Chapters 4 and 5).

Migration is intrinsic to all kinds of livelihoods, both biotic and abiotic, and acting at different scales. These dispersions, movements and shifts are the matter of intra-action and the matter from which matter arises. But not all migration is 'created equal'. Critiquing the displacement of livelihoods as a consequence and tool of power is a mainstay of postcolonial thought. As Rob Nixon suggests, critiquing Locke (and Jefferson), 'to dwell in movement is an unacceptable, uncivilized, irrational contradiction: you are improving neither the physical land nor yourself and, by extension, you're failing to advance the national interest. What counts as productive, legitimate, bureaucratically authenticated residence thereby becomes inextricable from the politics of visible self-improvement and the civilizational spectacle of the nation' (Nixon 2011: 165). These are not historical arguments, as attested to by the continuing construction of megadams and continuing displacement of livelihoods, and they are rehearsed as the basis for justification for the imperial (neo-imperial, or development) project.

Place, like power, is relational. Cultural geographers and philosophers of place have often generated notions of place in relation to space. In their widely cited introduction to *Space and Place: Theories of Identity and Location*, Erica Carter, James Donald and Judith Squires, for instance, suggest: 'How then does space become place? By being named; as the flows of power and negotiation of social relations are rendered in the concrete[!] form of architecture; and also, of course, by embodying the symbolic and imaginary investment of a population. Place is space to which meaning has been ascribed' (Carter, Donald and Squires 1993: xii). Here, place is privileged in opposition to space, as endowed with cultural meaning. Such comprehensions of the relationality of space and place, in their dualistic frameworks, can be and often have been useful for literary and cultural scholars in eliciting meaning-making processes of literary and cultural texts. However, as Doreen Massey poignantly argues in *For Space*,

> While place is claimed, or rejected, in these arguments in a startling variety of ways, there are often shared undergirding assumptions: of place as closed, coherent, integrated as authentic, as 'home', a secure retreat; of space as somehow originarily regionalised, as always-already divided up. ... [W]hat if we refuse that distinction, all too appealing it seems, between place (as meaningful, lived and everyday) and space (as what? the outside? the abstract? the meaningless)? (Massey 2006: 6)

For Massey, the foregrounding/backgrounding of space/place dualisms recedes behind a complexity of meanings present in space, giving rise to her idea of a 'simultaneity of stories-so-far' (cf. e.g. ibid.: 89). Rejecting such dualistic frameworks, in particular the comprehension of foreground/background, allows for the kinds of tangible knowledge metaphors that shift away from the privileging of sight as a mechanism of epistemic knowledge. This is productive for probing the materialities of concrete (and plastic, see Chapter 4). It also pushes back against the meaningful/meaningless associations of space that have engendered privileges of certain kinds of meanings, certain kinds of 'meaningful' places at the cost of such 'less meaningful' spaces, rendered concrete in the building of dams (and the inundation of lands).

I find it necessary to also 'dwell' shortly on the 'place' of displacement and the inherent normative constructions of inhabitation present in 'resettlement' that privilege the settled. If power is (also) created through disempowerment, I wonder if 'place' is also, or occasionally, created through displacement. This goes back to the issues of colonial encounters that rendered some peoples' places void on the erroneous and destructive assumption that engagements with place must entail settlement (or the building of (more than seasonal) dwellings). The denomination 'Development-Induced Displacement and Resettlement' (DIDR), on first glance at least, does not discursively address such peoples. Policies of 'resettlement' suggest, as indicated by the name itself, a previous form of habitation that can be recognized by 'settlement'.

In the introduction to this chapter, I noted the politics engendered by resistance to the planned construction of a hydroelectric power plant on the Franklin River in Tasmania. Another example, perhaps more cogent in an international context, is the UNESCO effort to translocate the temples of Abu Simbel in advance of the construction of the High Aswan Dam. In a press release to commemorate the fiftieth anniversary of the UNESCO project for the relocation of the Nubia Campaign, a rhetoric of 'international solidarity' is evoked ('50th Anniversary of Nubia Campaign'):[25] 'International expertise and funds were mobilized to dismantle and reassemble six groups of monuments in new locations … with the assistance of 40 technical missions from five continents' (ibid.). Former UNESCO Director-General Koïchiro Matsuura is quoted as describing it as a 'moving demonstration' of a 'miracle', a challenge to which the international community 'brilliantly rose' (ibid.).

The focus of this international effort was on the temples. The 'backstory' to the High Aswan Dam is one that entangles colonial past with sovereignty (in particular, the Suez Channel), with development, with nationalism and with livelihoods rendered dispensable. The High Aswan Dam is also prefigured by the particular locality of the Nile, which gave rise to certain agricultural entanglements that predate and prefigure the construction of megadams. The British colonial powers, in constructing the Low Aswan Dam at the turn of the nineteenth to twentieth century, systemized and intensified (i.e. localized more intensely) practices of harnessing flows of water which had been undertaken in the region for thousands of years. The reinforced concrete of the High Aswan Dam, then, both reiterates irrigation practices conducted in the region for thousands of years and also intensifies such practices. It is not irrigation itself that is new, but its scale, a scale enabled by the technologies of the material of concrete. The High Aswan Dam is thus not a 'foreign' technology superimposed onto the region. The scale of the dam, however, does suggest important shifts.

Hussein Fahim notes that Nubians were not the only people whose livelihoods were affected by the dam, and the temples of Abu Simbel were not only translocated to sites within Egypt, but were also removed to museums in the cities of the Global North. Crucially, 'thousands of nomadic peoples' who 'lived in the wilderness of desert habitat that flanked the narrow valley' (Fahim

---

[25] Although the UNESCO formed some fifteen years earlier (1945) in the aftermath of the Second World War, the 'Milestones' website notes the Nubia Campaign of 1960 to be the first, and (to date) largest, in a series of campaigns ('Milestones'). Cf. also '50th Anniversary of Nubia Campaign'.

1981: 57) were affected by its construction. 'One of the unfortunate concomitants of the inundation of Nubia', he notes,

> was that not only were the nomads unaccounted for when the potential impact of the project was assessed, but they apparently did not even receive warning or information relative to the change that was to occur and therefore had no advance notice as to the adjustments that would have to be made to accommodate the new situation. (ibid.: 90)

This begs the question: Can you dis*place* a nomadic people, whose relation to place is already dynamic and shifting? The acronym DIDR suggests not. Evidently some agencies, some lives and livelihoods, are privileged over others. This not only reiterates or re-references the displacement of power through the structures of the megadam – that is, that people (and other livelihoods) are shifted in the wake of the flood – but shows that some people are not *displaced* in the sense that they are not fully recognized by authorities as being *emplaced* in the terms of settled and sedentary livelihoods privileged by modern statehoods.

Rob Nixon uses the term 'unimagined communities' to articulate the ways in which such people are rendered invisible. Nixon's analysis, which explores notions of 'uninhabitants'[26] and 'development refugees' (an idea he traces to Thayer Scudder by way of Jacques Leslie), picks up most prominently on some essays by Arundhati Roy. The notion of 'unimagined community' (with its noticeable 'unarticulation', or absence, of Benedict Anderson's influential concept of imagined community) suggests both peoples who are imagined as not being part of the nation state or who are re-imagined as such. That is, either they are considered as always already outside the nation state, or, by function of some discursive or material in(ter)vention, reconsidered in these terms. 'If the idea of the modern nation-state is sustained by the production of imagined communities', Nixon argues, 'it also involves actively producing unimagined communities' (Nixon 2011: 150).[27] These 'surplus people' or 'ecosystem people' rely on the patterns of seasons and floods to sustain their livelihoods, which are becoming increasingly unreliable with climate change.

---

[26] This is a term that Nixon finds in Rebecca Solnit, and that Solnit draws from an interview with Janet Gordon (cf. Nixon 2011; Solnit 1999: 154–5), which is in turn predated, amongst others, by Terry Tempest Williams in her memoir *Refuge* (287). I think the idea has a lot of potential (I trace this term in more detail in Crane 2023b).

[27] The conditional is an addition. In the essay version (that preceded publication of these ideas in Nixon's monograph), this statement – the first sentence of that text – reads: 'The idea of the modern nation-state is sustained by the production of imagined communities but also by the active production of unimagined communities' (Nixon 2010: 62).

Nixon's concern is for people. A residual of the *material* is nevertheless traceable in his evocations of the 'literal' and 'metaphorical'. 'To erect a megadam', Nixon suggests, 'rendered *material* the trope of nation building', it 'was literally to *concretize* the postcolonial nation's modernity, prosperity, and autonomy' (ibid.: 166, emphases added).[28] Here, 'national self-assertion' becomes coupled with 'natural conquest' (ibid.), through uneasy shifting in registers between the ideas and the materials that are engaged and mobilized in order to enact such endeavours. Building on his analysis of Indian writer Arundhati Roy's essays, I suggest a way in which the materials, the concrete, of the megadam projects, elicit a concrete politics, one that recognizes the *concrete* effects and affects that assemble around concrete.

I shift, first, to Thomas King's (2012) collection *The Inconvenient Indian: A Curious Account of Native People in North America*. In the chapter 'One Name to Rule Them All', King sets the displacement of North American indigenous peoples for hydroelectric dam projects in a history of (settler) colonial control over Native Americans, First Nations and the Inuit of North America reaching back to first contact. King outlines initial trade and war relations, court cases (in the United States: *Johnson v. McIntosh*, *Cherokee v. Georgia*, *Worcester v. Georgia*), Canada's Indian Act, the flattening of differences by the linguistic determinant 'Indian'[29] and other acts of power as patterns of reiterated relocation through treaties and guises such as indigenous welfare. 'It wasn't until after World War II, as the economy began to heat up', King writes 'that Indians across North America found themselves being moved and relocated once again, this time to make way for large-scale industrial projects. In particular, hydroelectric projects' (King 2012: 95).

Whether considered as the repetition of spatialized exclusion, the colonial mechanisms of 'unimagined communities' or a material imperialism, King's tone is unmistakable:

---

[28] Citing Marq de Villiers's *Water: The Fate of Our Most Precious Resource*, Nixon points out that there were no dams over fifteen metres in height at the turn of the nineteenth/twentieth century, but at the turn of the twentieth/twenty-first century there were over 36,000. He reckons with two 'primary political contexts' behind these 'gargantuan structures': the Cold War and decolonization. With respect to the latter, 'the symbolism of epic dams' was seen by some leaders (he cites Nasser, Nkrumah and Nehru, in a rhetorical swoop of alliteration) where they could 'give *material solidity* to a newly acquired state of independence' (Nixon 2011: 166, emphasis added). Nixon notes the visual power promised by megadams 'as highly visible, spectacular statements that new nations were literally soaring toward development' (ibid.).

[29] In this chapter, I use 'Indian' in scare quotes to show my reliance on King's use of the term. Any collective term harbours the problematic potential to render a diverse multitude much more homogenous: within the United States, for instance, when 300 indigenous nations, tribes and peoples are rendered 'a minority' this radically transforms it into a 300:1 nation relation.

From British Columbia to Pennsylvania, from Saskatchewan to the Missouri River, from the Northwest Territories to Arizona, from Quebec to Washington, from Labrador to California, North America began building dams. Many of these dams were built on Indian land. The Army Corps of Engineers, in particular, was able to determine with amazing regularity that the best sites for dams just happened to be on Indian land. Even when there were more suitable non-Native sites available. (ibid.)

These are matters of ownership, matters of sovereignty, matters of matter, matters that matter.

Picking up on this theme later in this collection, King sardonically observes:

Treaty or no, I can't imagine that many folks in Washington really gave a damn whether or not Seneca land wound up on the bottom of a lake.

I know that's a rather cynical attitude, but, if you look at the history of dam building in North America, you might be surprised to discover how many excellent dam sites just happen to have been found on Indian land.

Then again, maybe you wouldn't. (ibid.: 232))

The displacement of North American 'Indians' elicits important transcontinental confluences with Subcontinental Indians,[30] in particular as rendered through specific political and material conditions.

Arundhati Roy's *Listening to Grasshoppers: Field Notes on Democracy* opens with an introduction with the subtitle 'Democracy's Failing Light'. In this subtitle, two issues are already linguistically intertwined: light in conjunction with democracy speaks to ideas of power, both political and pertaining to energy (electoral and electrical, I am tempted to write). In *An Ordinary Person's Guide to Empire*, Roy foregrounds notions of power more explicitly in an essay called 'Public Power in the Age of Empire', which addresses big dams explicitly. Power, in such titles, references the political power of governments, states, activists (and terrorists, however defined) and other agents and agencies; it also references electrical power, the power that runs our computers or lights (and, to a lesser degree, also other forms of power generation, i.e. fossil fuels running cars). Power is not additive here (meaning 1 + meaning 2), but rather can be mobilized

---

[30] Note, overemphasizing the point perhaps, that there is no one clear referent for either group of Indian/'Indian' – Hindu (Brahmin through to Shudras), Dalit, Muslim, Sikh, Tamil-speaking, Hindi-speaking, Bengali-speaking, Scheduled Tribes ... of the one Indian; and the Cherokee, Ojibway, Lakota, Navajo, Hopi, Choctaw, Inuit ... registered and unregistered tribes/nations/communities of the other.

for the different ways in which these (and other) meanings potentially premise, determine, conflate, extend and otherwise co-constitute each other.

Nixon's emphasis on the essays of Arundhati Roy throughout 'Unimagined Communities' works to elicit the incursive effects of the brevity of the form of her essays in contrast to the megalithic forms, not of the dams themselves, but of the technocratic documentation that accompanies them. The form of 'the agile personal essay' Nixon proposes 'was set against the ponderous, strategically impersonal epic report' staging 'intimate assaults on the calculated opacity, the profoundly consequential tedium, of the technocratic report that camouflages violence while clearing a path of it in a language scoured of emotion' (ibid.: 169). A pertinent analysis.

However, across the body of Roy's writings, something else is happening as well. Across the body of essays published on megadams, she repeats several points, sometimes verbatim. For instance, in the first essay of *An Ordinary Person's Guide to Empire*, Roy writes: 'The NBA [Narmada Bachao Andolan, the Save the Naramada Movement] believes that Big Dams are obsolete. It believes there are more democratic, more local, more economically viable and environmentally sustainable ways of generating electricity and managing water systems. It is demanding *more* modernity, not less. It is demanding *more* democracy, not less' (Roy 2006: 6, emphasis in original). In 'Peace is War', an essay included later in the same collection, she repeats this passage, almost verbatim.[31] This is arguably a function of the collection that brings together her writings and speeches which were dispersed through various channels. For an attentive reader, it does elicit an irritation. Thinking through this irritation, however, I have come to a number of conclusions that suggest the potentials for thinking through concrete materialities with this stylistic device. This kind of repetition – and there are more examples – draws attention to the similarities between the different contexts (and thus issues). It provides a 'way in' for those unfamiliar with the situation, that is, the repetition does not only alienate through irritation, but, at second consideration, might evoke also a sense of familiarity. It works away at the issue, a 'slow incursion' or 'slow erosion', similar to the way that water exerts wear on the concrete of the dam.

Repetition, further, is never 'sameness' or 'complete identity', it is, through the act of recognition, already enacting a small 'difference'. It 'defers', then, as

---

[31] Here, we read: 'The NBA believes that there are more local, more democratic, ecologically sustainable, economically viable ways of generating electricity and managing water systems. It is demanding more modernity, not less. More democracy, not less' (Roy 2006: 105).

well as 'differs', as it draws attention to the (impossibility) of the original (Jacques Derrida's *différance*). The material similarities between concrete forms somewhat outweigh the differences – otherwise it would be impossible to have a referent that we recognize under the term 'concrete' – but yet there are internal structural differences in concrete forms at a particular site through differences in aggregate, and, more obviously, from site to site (see also the introduction to this book). In this sense, Roy's essays are evocative of the materials, the concrete with its (local) water and aggregate and (global) cement and steel, shifting in scales and shifting weights. This has a radical potential to engender political action across spheres of similarity – of 'Indian' and Indian – without requiring sameness, or mimesis.

Roy's insistence on specific phrases then works to draw attention, in my reading, to the internal and regional differences whilst insisting on the structural/ material similarities. These are concrete similarities and differences that find a referent in concrete. Through Roy's essays then, and through the other texts, a political potentiality emerges, not only through the act of reading, but also through the potentialities of the materials themselves, to gather affects, which are rendered through and by the texts.

Cut

The divisive capacity of megadams to assemble responses is not only attested to through a number of the works I consider in this chapter, but it is also figured onsite. For example, on the wall of Matilija Dam, with an oversize pair of scissors painted onto its face with the dotted line suggesting 'cut here' (see Figure 3.3). The image, and the sentiment, has a precursor with the unfolding of the plastic schism on the surface of the Glen Canyon Dam by Earth First! activists. The incongruity of materialities – I'd guess the scissors would be less than apt at cutting through the thick walls of the concrete dam – combines with the agencies assembling around the viewer: The iconography 'asking' for a cut to be made.

The scissors can also invoke another cutting: the cutting of ribbons. New infrastructural projects invite celebration and commemoration (in a way that their maintenance does not). Scissors, indeed, are amongst the objects that the explosion mistaken as fireworks of Edward Abbey's 'Prologue' to his novel *The Monkey Wrench Gang* flings into the air: 'loose objects – gilded scissors, a monkey wrench, a couple of empty Cadillacs – slid down the appalling gradient … and launched themselves, turning slowly, into space' (Abbey 2006: 6).

Figure 3.3 'Matilija Dam, California 2014' by Carsten Meier. Copyright: Carsten Meier. Matilija Dam, California 2014. *DAM, Kerber, 2015*.

In *Encountering Development*, Arturo Escobar starts off with an epigraph, attributed to 'United Nations' (a body beyond bodies): 'Very few communities' we read 'are willing to pay the full price of economic progress' (as quoted in Escobar 1995: 3). Arundhati Roy, in the first essay in *Broken Republic* called 'Mr Chidambaram's War', writes: 'In our smoky, crowded cities, some people say, "So what? Someone has to pay the price of progress"' (Roy 2011: 2). This price – inevitably externalized – renders livelihoods expendable. And yet, if a community, a people, is unwilling to 'pay the (full) price', what is it that they are being sold?

Concrete manifests this quandary, well, rather concretely. In megadams, it acts as an index of development: its fixity, visibility, indeed the extent to which it constitutes a spectacle, all counter those traits deemed undesirable by the unflinching trajectory of development or modernity. Concrete manifests flow: the nomadic movements of the grounded but unlanded groups, forcing them to become sedentary, even sedimentary; the flows of capital, epitomized in the harnessed flow of water-becoming-power; and the power of modernity, especially in its indices of development, even as these are shifting elsewhere.

4

Unpacking plastic

Introduction: Unpacking plastic

One of the defining characteristics of plastic is its malleability, that it *can* be shaped. However, when we encounter it in day-to-day life, its form has been set. These characteristics hold also for our attitudes towards plastic: ostensibly malleable, but often set. The *plasticity* of responses to materials attests to the crucial ways in which cultural contexts frame material engagements. Heather Davis similarly observes: 'The plasticity that plastic embodies is an epistemic plasticity rather than an ontological plasticity. There is little plasticity in the actual objects of plastic', she suggests – though this depends on the item – 'but there is a lot of plasticity in our cultural investments in this object and the epistemologies that came to inform how it was made' (Davis 2022: 23). In this chapter, I trace the argument that plastic's capacity to be shaped into bottle caps which litter the coasts of the world's shorelines is *also* its capacity to be shaped into bags that hold IV liquids sterile in hospital environments. Further, plastic's capacity to preserve – its uses as containment against contaminants – extends beyond its (anthropocentric) usefulness, transforming into a capacity to preserve itself. The qualities that give rise to plastic conserving 'goods' (connoted as 'good') are the same qualities that lead to plastic's longevity as 'waste' (connoted as 'bad').

The story of plastic that emerged through Chapter 2 was somewhat one-dimensional. There, plastic was inevitably becoming-waste – if not already-waste – and wasteful. This is a story that is corroborated through many of the sources used in that chapter and reverberates further through documentaries, like Werne Boote's *Plastic Planet* and Ian Connacher's *Addicted to Plastic*, and through multiple advertising or environmental-awareness campaigns that stress the reduction of plastic. Such reactions, whilst crucial in drawing attention to the environmental impact of the proliferation of plastic on our planet, reduce the role of the material of plastic to a simplistic judgement: plastic is bad. However, the ways in which plastic shapes our lives are of a complexity that exceeds such dualistic accounts. This is not to say that plastic is inert – and as outlined in Chapter 2, research has shown it to be far from such – but rather to stress that the relations of plastic are multidimensional, shifting and porous. 'Solutions' to the plastic problem will also need to consider the complexity of these relations.

Susan Freinkel's account in *Plastic: A Toxic Love Story* flags such a relationship in its title, with the tension between toxic and love. She notes: 'Plastics draw on finite fossil fuels. They persist in the environment. They're suffused with harmful chemicals. They're accumulating in landfills. They're not being adequately recycled' (Freinkel 2011: 208). However, picking up on the 'love' trope established in the title, on the same page Freinkel also notes: 'In any event, it's not as if we can get a divorce. Plastics are one of the material foundations of modern life, and in many contexts, that's a good thing. We want our solar panels, bike helmets, pacemakers, bulletproof vests, fuel-efficient cars and airplanes, and, yes, even much of our plastic packaging' (ibid.). More than the plastic objects themselves, it is the practices elicited by, and with, plastic that shape and form attitudes.

Plastic practices, material interactions, find expression in myriad locations and activities. As noted in Chapter 2, the trope of listing plastic items ('concatenations of plastic') is widespread in the literature that examines plastic, in particular plastic waste. It also comprises one of the key scenarios of Werner Boote's *Plastic Planet*, where families from different locations were invited to gather all items of plastic from their households into a space outside of their homes for display: In these scenes, where Boote shows different households' reliance on plastic, families assemble with their items for a static shot within the documentary, much like posing for a photograph. There are not only several Western (probably Austrian or US American) households who partake in this display, but also a visibly poorer household in Kolkata, India. There the items are less numerous and voluminous than those of the wealthier (Western) households, and the items are also visibly well-used. This suggests a dynamic of plastic practices that stretches the timeline of the plastic as 'good' (as in 'a good' and as 'functional') beyond that of the Western households, as well as, crucially, a heightened exposure to the risks of plastics: Health risks increase as the physical integrity of the plastic items decreases, that is, with wear, tear and even cleaning, over the timeline of the items. Across Boote's scenarios, the lasting impression is one of livelihoods imbued with plastic. When this trope is mobilized to elicit negative feelings (of disgust, dismay, despair), as it often does, it neglects the particular ways in which plastic practices have led to particular ways of living in the world. In Chapter 3, I argued for the particular ways in which concrete manifested development, a conduit of modernity, through the generation of power. In this chapter, modernity finds a material conduit in the form of plastic.

The first plastic practice I examine is that of wrapping, which I distinguish from 'packaging' to shift away from industrial practices towards the more participatory practices in everyday life. I do so by examining cling-wrap, an item I presume most readers will be familiar with from their kitchens. On another level, I examine the artistic practices of Christo and Jeanne-Claude, whose wrapping is enacted at a larger scale. Where packaging has clearly become one of the 'bad' uses of plastic, the discussion of cling-wrap shows that the story of wrapping in plastic is more multifaceted. These findings are articulated further in conjunction with the large-scale wrapping practices of Christo and Jeanne-Claude, where I pay particular attention to the demarcation function of wrapping and how the plastic fabric they use in their work calls forth different interpretations of borders, in particular.

Cling-wrap, I argue, warps time. The temporal and spatial relations manifest in plastic (and other goods) have given rise to a trope of (philosophical) inquiry into what I call 'provenance stories'. Such stories attempt to trace the back-histories, or 'future histories' of materials and goods through time and space. They are, often, speculative in mode and work to elicit a fascination for 'things taken granted'. One example is James Marriott and Mika Minio-Paluello's sharing of ice cream at a conference in England and then tracing the story of the container back to Azerbaijan (Marriott and Minio-Paluello 2013); David Farrier similarly uses the moment of a plastic bottle falling to think through the deep-time provenances of oil in Pangaea (and its consequent journeys through the North Pacific gyre, cf. Farrier 2020: 105–15). Such 'provenance stories' work through temporal fascinations to reveal the (oftentimes deeply problematic) entanglements in logistics, consumption and marine exploitation (de Botton 2009) or violence and environmental degradation (Marriott and Minio-Paluello 2013), which these seemingly uncomplicated practices of consumption would otherwise mask. Provenance stories share conceptual grounds with the 'future artefact', even if the direction they take is different.

I turn to stories of medical settings for my second examples. Drawing on Doreen Massey's conceptualization of space as a 'simultaneity of stories-so-far', I tease out the various trajectories and stories that pass through two accounts of illness to elicit a more complex articulation of the patterns and practices which are entailed in plastic. Where Jody A. Roberts' account (2013) is explicit in its rendering of the ambivalence of emotional attributions to plastic (he calls himself a 'plastiphobe' in the subtitle), Mohsin Hamid's novel *How to Get Filthy Rich in Rising Asia* (also 2013), with its location outside of the 'West', renders the wonders of modernities in a different mode. Accordingly, this chapter, 'Unpacking plastic', works through various practices of packaging and packing – preservation, wrapping, transportation, sanitation and branding – to 'unpack' the various ambivalences and multivalences that accompany plastic in its various manifestations as a material of modernity.

Unpacking wrapping

I begin with practices of wrapping. Wrapping often functions to demarcate distinctions of value: The object that is wrapped is valued – even if, as in the case of cling-wrapping perishables, this is for a short duration – whereas the wrapping itself is disposable. Wrapping functions as barrier and as a practice that engages in definition: The materiality of the wrapping oscillates between invisibility and visibility. Particularly in the juxtaposition of cling-wrap practices and the wrapping practices of Christo and Jeanne-Claude, it becomes evident that plastic can be backgrounded (with cling-wrap, it is reduced to a function and is often transparent) or foregrounded (with Christo and Jeanne-Claude, it constitutes the practice itself, is constitutive of the artworks and must be seen).

Wrapping, importantly, is a plastic practice with a very short duration of interaction. Crucially, the temporal scales of the plastic material are incongruous with the temporality of the practice itself. When wrapping is disposed after its utilization, the aeons of time that pass in the compression of biotic matter into oil and the aeons of time that will come to pass in the biodegradation of the wrapping are compressed into a comparatively short duration of use. The expansive time and places of plastic are condensed, and then thrown 'away', with the discrete, isolated and isolating, materials of wrapping.

I first trace the tangible wrapping practices of (some)[1] quotidian lives using the example of cling-wrap, as evinced and pre-empted through a television advertisement from its early days (the 1950s, in the United States). In stressing

---

[1] This 'some' is deliberately vague: It will map to cover me and most (probably all) of my readers, and the upper, middle and upwardly mobile classes across the globe, as well as the poorer classes in the Global North – with some, important, interruptions, including those who refuse plastic wrapping. There is a whole genre of 'plastic-free' guides, online and printed, which I do not examine in this chapter, but should prove interesting for a critique in the vein outlined here.

the idea of *wrapping* rather than the broader category of packaging, agency is shifted to the individual rather than somewhat obliquely relegated to an intangible corporate entity. I stress engagement in the practice itself, as is the case with cling-wrap, or the reception of the (artistic) intervention into spaces and environments (or, alternatively, the documentation thereof).[2]

Unpacking wrapping I: Cling-wrap time-warp

With cling-wrap, Roland Barthes' description of plastic as 'the first magical substance which consents to be prosaic' (Barthes 2000: 98) comes into its own. Early television advertisements to introduce cling-wrap into the market worked to prove its positive characteristics by emphasizing its utility and practicality over other (established) materials used for wrapping food. I have chosen one advertisement, a 1953 advertisement for Saran Wrap (a Dow Chemical Company product), to demonstrate the way cling-wrap (as practice, as idea) was introduced to the market.[3] The various selling points elicit an understanding of plastic practices and meanings generated during the post-war US American obsession with plastic. Close attention to historical and socio-geographical circumstances both 'on-' and 'off-screen' reveal cling-wrap's far-reaching entanglements.

This television advertisement for Saran Wrap runs for 54 seconds and appears to be directed at a US market (an assumption I derive from the language and accents, as well as the location of the Dow Chemical Company in the United States). The visuals of the actress's dress and, importantly, the setting – the style, size and set-up of the kitchen displayed in the background – seem to interpellate a market segment comprised of middle-class housewives. The actress, performing the role of the housewife in the advertisement, is white and female, assuming norms attributed to US suburban middle-class demographics of the time. This impression is reinforced by repeated references to thrift, which I address in more detail later.

The use of a male voice-over at the beginning and end of the advertisement negotiates the transition into and out of the advertisement and echoes the

---

[2] For another project, I recently found myself taking a photograph of a supermarket refrigerator filled with (meat in) plastic packaging. Perhaps I will pursue this image idea another time.
[3] See https://archive.org/details/1953CommercialForSaran-wrapad2. Timestamps are omitted in the discussion here, for reasons of clarity (readability) and also due to the fickle nature of internet video platforms. Note that many consequent advertisements tend to focus on brand establishment, that is, on suggesting positive qualities associated with a particular brand, rather than with the applications and qualities of the material in general.

transition into and out of the public sphere. These are gendered spheres: The male voice occupies a position of authority, being able to transverse these different spheres as well as the bounds of the message itself. This impression of buffering the private (coded: female) into the public (coded: male) is reinforced by the visual references to the supermarket, a public space – which are shown in conjunction with the male voice-over at the beginning of the advertisement – and to the logo of the Dow Chemical Company, which occurs, again, in the public sphere, at the end. Within this framework, the bulk of the advertisement takes place in the (presumed) feminine and domesticated space of the private kitchen.

The main selling points of cling-wrap proposed in the advertisement are the (visual) clarity of plastic, its clinginess, the extended freshness and tastiness of food in storage the product confers and convenience of time planning it enables. Each of these characteristics has implications for the material itself and for its position in a wider context and warrants further explication.

The clarity of the plastic wrap is established both verbally and visually in the advertisement. The actress declares that 'it keeps everything you wrap in plain sight' whilst holding a portion of the translucent and as-yet-unwrapped plastic between her face and the camera, also obliquely suggestive of the veil as a symbol of feminine purity (or virginity). In moves that echo the introduction of many other application of plastics, cling-wrap *displaces* other materials by emphasizing plastic's material advantages: here, the advantage is its translucency. In explicit contradistinction to paper or tinfoil, cling-wrap preserves the visual presence of the items thus wrapped, because we can see through it.

However, whilst plastic, through its clarity, maintains visual access to the wrapped goods, it is accompanied by a reduction of access to the wrapped items with respect to other senses. Specifically, the more affective (or immediate) modes of interaction – such as the tactile, gustatory or olfactory – are reduced through the barrier of plastic wrap. Cling-wrap thus reduces the physical possibilities of engaging with food through touch, smelling and tasting, at the same time as preserving the distanced control offered through vision. The material practice of preservation through plastic works to cordon off capacities for bodies to interact or engage with the stuff that is preserved: Cling-wrap is not simply a border, the materials of packaging and wrapping themselves emerge as conduits of forms of (re)presentation and preservation.

That is, the plastic barrier emerges as integral to the *preservation* of artefacts *as* objects. This is an effect of all manner of wrapping practices. Things that are wrapped are objectified and contained; plastic wrapping amplifies this practice along a divide in sensual interactions, reinforcing the visual sense and – perhaps *by* – reducing the affective and tactile senses. Plastic is particularly good at this. As Heather Davis suggests, 'The aesthetic effects – as in *aisthesis*, or affects produced by our sensorial experience of the environment – have been entirely re-ordered by the presence of plastic' (Davis 2015: 348). Only the cling-wrap's shiny veneer renders a visual presence of a wrapping material, emphasizing the object *as* object.

If the materiality of cling-wrap recedes, becoming almost imperceptible, once it has been put into use, anyone who has used cling-wrap will recognize how it is incontrovertibly stubborn in its application. In the Saran Wrap advertisement, clinginess is one of the main selling points. The advertisement suggests this results in a simplification of process. Other (auxiliary) materials, such as strings or rubber bands used to hold down the wrapping materials, are thus rendered superfluous. However, in my experience, it is exactly this quality of cling-wrap that oftentimes makes it a frustrating practice, as the cling-wrap becomes wrapped in itself. Attempting to preserve foodstuffs in the kitchen with cling-wrap is an apt metaphor for, and practice of, the entanglements of daily lives in the materials we interact with.

Cling-wraps also cling, metaphorically, to their histories (in particular: of applications and production). Whilst Mary Bellis suggests that Saran Wrap 'might be the first plastic wrap designed specifically for food products' (Bellis 2017: web), other sources suggest that predecessors of the plastics utilized in cling-wrap products encompassed various other applications. These applications are broad, ranging from use in fighter planes as protection against salty spray through to upholstery coverings. Following the end of the Second World War, Susan Freinkel notes, 'The vinyl-based compound Saran, which had proved so useful in protecting military cargo, was redeployed to the short-term protection of leftovers' (Freinkel 2011: 121). Alternatively, Stephen Fenichell has suggested that Saran was originally envisioned as a chewing gum yielding theatre-seat covering (cf. Fenichell 1996: 211). Today we encounter cling-wrap in the kitchen, or perhaps in at the hairdressers, or as wrapping for suitcases at airports (a somewhat mysterious and superfluous practice that leaves me baffled).[4] It is used to dehydrate bodies in preparation for beauty pageants and also in scientific experiments, for instance, wrapping biological matter for preservation

---

[4] Tom Fisher suggests that 'sealing the suitcase in plastic ameliorates passengers' fear for the security of possessions that disappear into the baggage handling system. The visual effect of wrapping a suitcase in many layers of plastic film may also be relevant, as the film projects its characteristically soft but glassy sparkle to other passengers and, presumably, to baggage handling staff' (Fisher 2013: 118, fn 6).

in the laboratory. Cling-wrap has, since being invented, also found many new applications invented for its use.

The use of Saran Wrap – which used polyvinylidene chloride, or PVDC – to wrap biological matter in laboratory experiments was later linked to the development of tumours in the kidneys of rats that were wrapped in the material (cf. Freinkel 2011: 262). After SC Johnson acquired the Saran brand from Dow Chemicals in the late 1990s, PVDC was removed from the production process, citing some concerns about its safety (cf. Bellis 2017: web; for more on toxicity and plastic, see Chapter 2).[5]

But not all cling-wraps use or have used PVDC. Glad Wrap, for instance, uses a polyethylene, which accounts for more than a third of all plastics produced and sold worldwide and is considered the first commodity plastic (i.e. it is cheap and abundant) (cf. Freinkel 2011: 61, 236). Glad Wrap has its own, specific, 'clingy' history: Launched by Union Carbide in the 1963, its history must be thought together with the Bhopal disaster of 1984. Wil Lepkowski cites a 'carbide community relations official' interviewed in 1991 as stating that before Bhopal, 'the whole chemical industry operated on the basic assumption that what we did within our fences was none of anyone's business … [.] And the people outside the fences didn't think it was any business of theirs, either. Bhopal changed all that' (as quoted in Lepkowski 1994: 25). Befitting of a company that produces products that forge boundaries and create containers, compartmentalization was an industry-wide strategy. The later divestment of Glad by Union Carbide

---

[5] Mary Bellis notes that the discovery of polyvinylidene chloride (or PVDC) was accidental, accredited to Ralph Wiley, a student doing cleaning work in the Dow Chemical laboratories in the early 1930s, who found remnants of a 'failed' experiment. The intentionality of 'designed specifically for food products' (Bellis 2017: web) is undermined by an 'accidental discovery' story familiar to many applications of plastic (cf. also the explication of invention and discovery in the introduction). Polyethylene, too, was invented in unpredictable circumstances. In retelling this story, Freinkel uses the term 'discover' but is also quick to point out its experimental-cum-accidental circumstances:

> Polyethylene was discovered[!] in 1933 by two chemists at Britain's Imperial Chemical Industries who were noodling[!] around in the lab exploring how ethylene reacted under high pressure. In a series of experiments – including one that blew their reactor and much of their lab to smithereens – they found that with extreme pressure and the catalytic cajoling of benzaldehyde and a bit of oxygen, ethylene molecules hooked together into chains of stupendous length. The flakes of snow-white waxy stuff they found at the bottom of the reactor vessel were 'so unlike polymers known at the time … no one could envisage a use for it', one of the researchers recalled. (Freinkel 2011: 60–1, ellipsis in original)

In this narrative, 'discovery' belies the notion of intentionality; the use of the informal verb 'to noodle' suggests this, as well as the initial incapacity of the present scientists to 'envisage a use' for the 'waxy stuff' they 'found' (i.e. it was not 'produced'). Application does not drive experimentation, nor does it derive from it: Taking considerations of material agency into account, it seems more appropriate to suggest that the applications of the plastic emerged *with* human interactions, in particular those by people who were open to the ways in which the plastic polymers attested qualities of the unexpected kind.

can be seen as a corporate attempt to free a brand name from the liabilities that cling to the disaster. Cling-wrap's seeming innocuousness as a quotidian addition to the kitchen thus emerges as a complex entanglement in histories of experimentation and disaster. The histories of various cling-wrap brands suggest entangled relations of the material (or product) that stretch through the public sphere.

Temporal considerations are key to understanding another selling point in the 1953 advertisement, namely, the saving of time and the saving of food. Saran Wrap, the actress claims, provides convenience in time planning, it is 'a time saver as well as a food saver'. However, the issue of 'saving time' broached in the advertisement is not as straightforward as the phrase implies. Strictly speaking, no time is actually saved in the preparation of lunches. Time is rather diverted, or made more plannable, that is, overall preparation time is not reduced, but brought forward. In this respect, cling-wrap might be considered more fittingly as a material of 'storing time' than 'saving time'.

Elizabeth Shove suggests the concept of 'storing time' to cover technologies and devices that offer such forms of 'hypermodern' convenience' (Shove 2003: 171). Her list comprises electric devices, like microwaves and freezers, as well as telecommunication devices, to which the comparatively low-tech device of cling-wrap might be added. 'In allowing users to "store" time, defer activity or manage and minimize interruption', Elizabeth Shove argues, 'tools of this kind enhance capacity for autonomous organization' (ibid.: 172). I want to think about how time is not necessarily saved in practices of pre-preparation, but becomes manageable through the allocating of activities to specific time periods.

In the case of leftovers, where the successful storage of post-preparation food *can* allow for later consumption without the time invested in food preparation, thriftiness does result in a 'saving' of time (and a reduction in food waste). In this application, the practice of wrapping in plastic delays decay by creating a barrier to decomposing bacteria. Saving food is folded into a temporal dimension by the creation of borders between different kinds of organic matter, that is, food and bacteria, visualized by the folds in the cling-wrap image in the advertisement itself.

By extending the time periods through which food is considered palatable, cling-wrap and similar products push back on organic time frames. Further, as Gay Hawkins, Emily Potter and Kane Race suggest, 'The value of Shove and colleagues' approach lies in its insistence on the role of materials as actants that can suggest and transform practices – that is, on practices as complex assemblages of human and nonhuman that are always on the move' (Hawkins,

Potter and Race 2015: 84). Cling-wrap's effectiveness is codified through its performance of shifts in temporalities by the creation of borders. In this sense, time is moved by specific materialities of cling-wrap: cling-wrap warps time. But not only time. If we accept Elizabeth Fisher's proposal that the first tool was not a hammer or knife but rather a receptacle for moving things about, we might recognize, like David Farrier, that the 'invention of the vessel broke open time and space' (Farrier 2020: 91). As Farrier suggests, 'Once our ancestors could place items they might need later in a bag, they could satisfy their hunger when and where they chose. They no longer needed to go to the river to drink or the bush to eat; they could carry the river or the forest with them' (ibid.). And whilst cling-wrap is not strictly speaking a receptacle, its function of shifting relations to time is also a function of shifting relations to space.

There is another spatio-temporal dimension at work, beyond the 'edges' of the commercial: I refer to the long perseverance of the plastic as material. It cannot really be surprising that a material employed for preservation against organic decay works to preserve itself. The stability of plastic as object is predicated on a dramatic compression of time. The daily use of cling-wrap in households is one of the most remarkable instances of this compression: Its 'utility' or duration as a useful household object extends from the moment of unwinding some wrap from the roll to the time of discarding, a period which rarely extends beyond a few days. Of course, when the practice of unwinding from the roll is less successful, as is my occasional experience, the span of utility is compressed into a matter of seconds: it's almost over before it has started. This duration of engagement is nestled within several other time spans: the duration of 'proximity', from purchase of the roll of cling-wrap to the moment waste, is removed from the household; in turn nestled into the duration of 'form', from the moment of production to the moment it enters landfill (or recycling plants);[6] and, most dramatically, the duration of 'material', from the formation of the fossil fuels from which the plastic is won to the potential future of the polymer chains finally decomposing. These nestling functions – utility, proximity, form, material – are predicated, finally, on some kind of 'elsewhere', an externalized, expendable place, where the plastics do their final work. This final timeline is, as Max Liboiron argues, at best a guess:

---

[6] Hawkins, Potter and Race suggest that 'turning recovered bottles into a resource for diverse forms of plastic manufacture involves executing distinct practices of destruction' (Hawkins, Potter and Race 2015: 131). They call this process 'disassembly' and note how recycling requires the reversal of many of the affordances of the plastic bottle (their focus), for example, the once unbreakable and uncrushable qualities of the bottle emerge, in recycling, as powerful constraints to the transformation of the material into pellets (cf. ibid.: 132).

You may have noticed that temporal estimates of plastics breaking down (one thousand years for this kind of plastic, ten thousand for this other kind) exceed the amount of time that plastics have existed. Most of these estimates are modeled from data created in labs (in UV-saturated, vibrating, acidic set-ups that rarely mimic actually existing environmental conditions) and are based on the idea that the rate of weakening polymer bounds will proceed on a regular curve. They do not anticipate the effects of metabolites or the molecular chains that polymers might break into. They cannot anticipate how future environmental relations will absorb, adapt to, and otherwise influence these rates of breakdown or the effects of many types of plastics in diverse environments over long periods. (Liboiron 2021: 17, fn 64)

And so, again, cling-wrap warps time. In its many applications developed to resist bacterial or other organic processes of interference in the structural integrity of the material, it should really be no surprise that plastic perseveres in the environment. Its permanence as a material exceeds its permanence as object (or as 'useful' objects for preservation). The persistent presence of plastic in the oceans of the world and in the bloodstreams and endocrinal flows of its (organic) inhabitants reveals plastic's capacity for *self*-preservation. The quality that ensures its 'usefulness' is the same quality that ensures its 'wastefulness'. To push this point further, cling-wrap reduces waste by always being on the cusp of becoming waste.

The 1953 advertisement for Saran Wrap presents (a particular brand of) cling-wrap to 'the world'. The advertisement highlights the ways in which the plastic practices of cling-wrap relate to time (saving/storing time) and to preservation and explicates its visual clarity and its clinginess. These selling points, in turn, elicit a greater comprehension of the entanglements of wrapping as a quotidian practice in larger frameworks of materiality, time and (geopolitical) histories. My use of cling-wrap folds these multiple dimensions of history and practice into it, making for a particularly sticky entanglement.

Unpacking wrapping II: Christo and Jeanne-Claude, presenting, preserving

Practices of wrapping are also used in acts of giving. In such practices, the present is marked as – perhaps even becomes – a present through the wrapping: The immediate contextualization, which is also an immediate containment, acts to bestow a significance on the object which is wrapped. The idea that presentation is a crucial component of semiotic processes of bestowing of meaning is also readily apparent in modern art practices: The context of the art gallery or museum gives rise to certain interpretative and curator practices, whereby mundane materialities become marvellous mysteries.

I turn now to aesthetic practices of wrapping, specifically the wrapping projects of Christo and Jeanne-Claude. Werner Spies, centring Christo (and leaving out Jeanne-Claude), suggests that 'unlike the world of advertising from which Pop Art took its cues, Christo shifted the focus from packaging to contents' (as quoted in Baal-Teshuva 2016: 35–6). As Christo asserted, the fabric is '*always* a very essential element of our works' (in interview with Nathan 2015, emphasis in original). As one, if not *the*, defining characteristic of Christo and Jeanne-Claude's oeuvre, their wrapping warrants attention in this chapter for the ways in which the materiality of the plastic fabric configures presentation and preservation through the large-scale installations and interventions of the artists.

Wrapping practices – all packaging practices – function by means of creating a border between object and surrounds. To disregard the materiality of this border is to jettison the various social and cultural functions of wrapping and packaging practices that come into being *through* these practices. Further, the materiality cannot be divorced from the spatial and temporal dimensions of the wrapping (or packaging) itself. With plastic, such dimensions are of a non-human timescale. Turning from cling-wrap to Christo and Jeanne-Claude's installations entails a shift in scale and dimension and also constitutes a continuation of my analysis of the transient effects of plastic packaging.

Jacob Baal-Teshuva describes the *Surrounded Islands, Biscayne Bay, Greater Miami, Florida, 1980–83* project as 'one of the most spectacular projects conceived and created by Christo and Jeanne-Claude'.[7] Wrapping islands, or surrounding them, picks up several themes from previous chapters, including the flow of water. This example of wrapping (see Figure 4.1) is a function of a very particular site – Miami – and as such brings forth associations of incursions into natural spaces that pervade the (geographical) context of Florida: the draining of swamps to reclaim real estate, for example, which displaces land and water in order to provide habitats more tenable to land development.

---

[7] My copy of Jacob Baal-Teshuva's *Christo and Jeanne-Claude* arrived, either aptly or mindlessly, wrapped in a sheath of cling-wrap (forging a further connection between this section and the previous one). The book was thus *presented* to me as a particular kind of artefact (not all books are wrapped thus, but art volumes and other large-format glossy books – I call them 'coffee-table books' – often are).

Figure 4.1 'Surrounded Islands, Biscayne Bay, Greater Miami, Florida, 1980–83' by Christo and Jeanne-Claude.
Source: Photograph by Wolfgang Volz. Copyright Christo and Jeanne-Claude Foundation.

Baal-Teshuva's description of the project alludes to such practices when he states: 'This mammoth project – accommodated to the natural environment, symbiotically blending with it without harm or disturbance – was a work of great beauty, delicacy, and poetry, as well as daring, and posed considerable risk and technical difficulty' (Baal-Teshuva 2016: 55). The image here is a photograph of the realized project.[8] The project used some 603,870 square metres of bright pink woven polypropylene fabric to surround eleven islands in the Biscayne Bay, Florida. Each of the islands was circumvallated by a ring of fabric 61 metres (200 feet) wide. The effect is rather dramatic: The contrast between the lurid pink of the plastic fabric and the blues and greens of the surrounds is striking, and the scale of the project immense.

The installation was in place for two weeks in May 1983 and construction and monitoring involved some 430 workers. The marine and land crews working to prepare the islands for the installation 'picked up debris from the eleven

---

[8] I have included an image which can also be found on the artists' official website, where it is possible to navigate through many of their projects, including collage sketches as well as photographs for both realized and unrealized projects. Christo and Jeanne-Claude created numerous sketches and collages, which were sold in advance of the various projects' realization as a means of raising funds. This practice could be seen as a sort of protean 'kickstarter'.

islands, putting refuse in bags and carting it away' and 'removed some forty tons of varied garbage: refrigerator doors, tires, kitchen sinks, mattresses, and an abandoned boat' (Volz and Pachnicke 2013: 167): This emphasis on the removal of rubbish seeks to frame the project's positive environmental impact – however, other ecological and geological processes were interrupted for a fortnight. The transiency of their installations is emphasized in Baal-Teshuva's introduction to the works of Christo and Jeanne-Claude: 'After two weeks all sites are restored to their original condition and the materials recycled' (Baal-Teshuva 2016: 9). Project descriptions included in the appendix to another volume (edited by Wolfgang Volz and Peter Pachnicke) also note the final destination of the materials used in the projects themselves (whether recycled, and by whom, or to whom the materials were given). Great attention is thus paid to the material impacts of the installations beyond their short duration (usually around a fortnight), both by the artists themselves and by the authors of such catalogues. This aspect of the descriptions of the installations emphasizes the interactions between the artistic and *material* legacy of the artworks.

Many of Christo and Jeanne-Claude's artworks are made of plastic fabric installed *as* and/or installed *in* liminal spaces. The fabric, as wrapping, creates a border between that which is wrapped and its environment. At the same time, the plastic fabric oscillates between (lurid) presence *as* material border and (seeming) absence as functional border. The plastic fabric is present *and* absent, it is object *and* function, it is subject *and* context. By emphasizing the distinction between the object and environment, the tenuity of the border itself is also stressed: That is to say, as temporal, even tenuous, artificial border, the plastic emphasizes the artifice of borders in general.

The ephemeral quality of the plastic fabric seems a strange choice for the installation of such borders or walls; however, this is the very quality for which Christo suggested the fabric was chosen, for example, in the *Running Fence* project, which was completed in California in 1976.[9] As he stressed on numerous occasions, the fabric is a crucial component of their oeuvre: '*always* a very essential element of our works was the cloth, the fabric, which was the principle material used to translate this nomadic, temporary

---

[9] As Christo has explained, 'The first *Running Fence* idea was to have a fence along the Berlin Wall, and there are some drawings from between 1970–1972 of a fabric fence in West Berlin, hiding the wall … that was my first project for Berlin' (Christo, as quoted in Baal-Teshuva 2016: 78, omissions in original). That wall, as noted in my introduction, has been transformed from political boundary to a site of cultural memory. Under the Trump presidency, in particular, southern US American states, including California, emerged as the site of a different, and yet similarly politically instrumentalized, wall.

existence of the work' (Christo, in interview with Nathan 2015, emphasis in original). The impermanence of the fabric thus exudes into the artwork and unsettles the (ontological status of the) borders constituted by (and critiqued through) the artwork.

Christo and Jeanne-Claude not only used fabric to constitute borders, they also used it to *wrap* borders or objects that broach borders, for example, river banks, beaches and bridges. In such artworks, the plastic fabric emphasizes the functionality of borders themselves by delineating these borders from their environments. The wrapping of bridges, for instance, draws attention to the boundary-demarcating function of the fabric (plastic) at the same time as to the boundary-broaching function of the bridge. Function and form work to reinforce each other, as in the wrapping of bridges in Paris, *The Pont Neuf Wrapped* (1975–85).

In addition to the *Surrounded Islands* project, further examples of 'border wrapping' include the beach and cliffs in the *Wrapped Coast* project installed near Sydney in 1969 (this installation was left standing for longer than other projects, 10 weeks in total); a similar, but smaller-scale, project called *Ocean Front* in Newport, Rhode Island, from 1974; and, more recently, the *Floating Piers* project (Figure 4.2, and analysis later), also installed on the cusp of water and land, realized in the summer of 2016, on Lake Iseo, Italy.

In more recent permutations, in particular *Surrounded Islands* and *Floating Piers*, the plastic fabric conceals that which it wraps, beneath lurid pink and orange, respectively; in the earlier projects (*Wrapped Coast* and *Ocean Front*), the loosely woven fabric allows shadows of the liminal space to remain visible. In some projects, the fabric appears translucent; in others, its lurid colours draws attention to the wrapping, both as noun and as verb. Erika Doss suggests that 'for 50 years, Christo and Jeanne-Claude organized temporary and massively scaled public projects that centered on the manipulation of space and the disruption of conventional, or familiar, views of places, objects, and ideas' (Doss 2017: 199). Regardless of whether the plastic was either brightly coloured or not, the act of temporarily wrapping draws attention to a cusp site: Liminality is performed or practiced in both spatial and temporal dimensions. It is through the very plasticity of the fabric that this effect is achieved: The plastic, in Christo and Jeanne Claude's artwork, makes the environment 'plastic'. Tangible, but also transient.

The conception of these projects was inevitably lengthy, far exceeding the duration of the actual installations. The sketches and collages composed in advance of the installations have 'durations', or legacies, as artworks that

Figure 4.2 'The Floating Piers, Lake Iseo, Italy, 2014–16' by Christo and Jeanne-Claude. *Source:* Photograph by Wolfgang Volz. Copyright Christo and Jeanne-Claude Foundation.

exceed the installation itself, as do the (documentation) photographs and other impressions made during the installations themselves. Similarly, the 'legacy' of the fabric far exceeds its use as a wrapping – as I have also argued in connection with cling-wrap earlier. This encompasses not only the 'duration' of the fabric *as* fabric, that is, as plastic in a particular form, but, again, the long 'duration' of the plastic that stretches back to the fossils that fuel its production (both as energy source and as construction material) and to the anticipated 'duration' of the altered fossils as plastic in the future tense, that is, as future artefacts.

Floating Piers entailed the installation of bright orange piers, bridges and walkways, on Lake Iseo, connecting – for the duration of the installation – the island San Paolo with the mainland. The project entailed the installation of anchors, a floating dock system (made of polyethylene cubes), as well as the visible nylon fabric. Reports suggest that over 1 million people visited the installation in two weeks it was open to the public. In her interpretation of the time frames and evocations of impermanence, Erika Doss adopts a critical stance of this particular project, pointing out that *Floating Piers* installation on Lake Iseo in the summer of 2016 not only entailed the direct involvement of the Beretta family – weapons manufacturers who own the San Paolo island which was surrounded by the orange plastic piers, and who were instrumental in facilitating numerous permits required for the installation – but also was haunted by different, contemporaneous presences in the Mediterranean, only some 200 kilometres away: specifically, refugees (cf. Doss 2017), those who could not 'walk on water' to the shores of Italy (and the EU), but who instead drowned. The plastic, forging connections between islands, was afforded an agency and role that some humans were not. The piers, in this analysis, give way to drawbridges, ready to be raised in defence of 'fortress Europe'.

The wrapping projects occupy a space of tension between the everyday experience of wrapping and concealment/preservation and the extraordinary scales and impacts of Christo and Jeanne-Claude's artworks. Their use of (plastic) fabric was integral to their oeuvre – even beyond the wrapping projects: David Bourdon has suggested that they 'touch the world with wonder. From those modest beginnings in Paris they have gone on, over a career of fifty years, to wrap everything' (David Bourdon, as quoted in Baal-Teshuva 2016: 16). In the works of Christo and Jeanne-Claude, the plastic fabric wrap is not *simply* border, and the ways in which the materials of packaging themselves are conduits of forms of (re)presentation are emphasized. The wrapping itself becomes an artefact that pushes back on the distinctions of framework/work or paratext/text, oscillates between form and functionality, and foregrounds its very materiality, with all the temporal and spatial ramifications of worldly presence.

Far from a mundane practice, by no means an innocuous presence, plastic wrapping *reveals* as it *conceals*. The far-reaching entanglements of the quotidian practices of cling-wrap, the chemical, historical and environmental/political consequences, are not divorced from the materiality of the plastic, but enacted with them. The large-scale installations of 'wrapping' realized through the works of Christo and Jeanne-Claude perform and celebrate the functions and forms of wrapping, foregrounding the otherwise potentially mundane and innocuous. Even as it constitutes borders, plastic materiality flouts categorical stability.

Tracking plastic: Provenance stories and material trajectories

Materials forge relations through their movements in spaces and times, by coalescing in specific forms and, crucially, through the compressions that transform them into a speculative dimension of materials-becoming(-waste).[10] This has a temporal dimension, the long duration of the pre- and after-life of

---

[10] Plastiglomerate is a good example for this, see 'Introduction', Chapter 2 and 'Conclusion'.

the future artefact, as argued earlier with respect to the varying durations of cling-wrap plastics. It also has spatial dimensions, as James Marriott and Mika Minio-Paluello's tracking an ice-cream container as an 'origin geography' (Marriott and Minio-Paluello 2013: 172) suggests. I use a slightly different terminology: 'provenance story'. Whilst 'provenance' generally suggests the 'origins' of something and thus maps nicely onto Marriott and Minio-Paluello's term, the Latin roots of the word 'provenance', *pro-* 'forth' and *venire* 'come', might also be read against this grain to allude to what *will* come forth, that is, the dimension of a (speculative) future of the materials as future artefact. 'Provenance' also reverberates with 'inventory', through the shared roots in *venire*, a 'coming into relations' with and through materiality.

Marriott and Minio-Paluello start their chapter 'Where Does This Stuff Come From? Oil, Plastic and the Distribution of Violence' with a narration of shared ice cream (at a conference), after which they direct their audience's attention to the story of the ice-cream container. The ice-cream container is followed back along its line of production, tracing a 'destructive "pre-life" of plastics' (ibid.: 172). The narrative they then develop works as a corrective to what they suggest is a 'deeply ingrained' practice, namely of 'considering plastic objects "as they are", rather than as the products of a non-plastic substance' (ibid.). In the case of the ice-cream container, as with many plastics, this is a story that begins with oil.

Marriott and Minio-Paluello thus compose a narrative of the material of plastic objects before these objects are formed. An approach attending to materiality, then, rather than objects or things. Where others have rather demonstrated the 'destructive after-life of plastics' (ibid.), they 'follow the passage of that material from oil-bearing rocks, through drilling rigs, pipelines, terminals, depots, refineries, factories, distribution centres and shops, to homes' (ibid.). The ice-cream container, no longer contained, thus covers many thousands of kilometres (and many thousands of years) before becoming packaging, a disposable item. As oil, they suggest, it is drilled in Azerbaijan, before traversing Georgia, Turkey, Italy and Austria. It is refined and made into resin in Germany, transformed into containers, which are then transported to Gloucester, filled with ice cream and distributed throughout England. Marriott and Minio-Paluello invite their audience and, later, readers, to imagine the materiality of the ice-cream container extending temporally and spatially back along this trajectory. Their emphasis on the political dimensions – evinced by referencing 'forms of exclusion and violence' – offers an important corrective to the seeming innocuousness of the ice-cream container.

Beside their attention to specificities, theirs is a speculative narrative:

The exact passage of Clyrell EC340R to Sainsbury's in New Cross is relatively opaque to us. We cannot be 100 per cent sure that there is Azeri oil in these specific boxes – once again, the journey has entered a 'forbidden zone', though this time it is the globalized processes of manufacture and distribution, rather than forms of exclusion and violence [as was the case of the pipelines for the crude oil earlier in the narrative], that are obscured. (ibid.: 180)

Consumers are thus brought into (narrative, speculative) relation with the relations that bring the object into being. As material, the ice-cream container bears relations that exceed its form; its plasticity forges connections that stretch beyond its presence in the conference room across time and space. Tracing the 'origins' of the container in this way elicits a sense of the stretchiness of entangled relations which produce, and are condensed in, everyday objects.

A similar trope is mobilized by Alain de Botton in the chapter 'Logistics' from *The Pleasures and Sorrows of Work*, where he examines the technologies and movements entailed in eight-year-old Sam's dinner of tuna. The tuna, eaten in central England, is traced to the Maldives in the written text as well as an accompanying photo-essay ('A Logistical Journey'), with photographs taken by Richard Baker. The journey itself is initiated by 'a claim as concise and tantalizing as an epitaph on a gravestone' (de Botton 2009: 46) written on the (probably plastic) wrapper, reading 'Caught by line in the Maldives' (ibid.). Whilst the choice of object (the tuna) is framed as being somewhat arbitrary in this piece, the tuna does offer a telling metaphor, revealing much about the temporal dimension reversed by the telling of the tale (the photo-essay is actually chronological). De Botton writes:

> The tuna's lessons, while played out in particularities, are nonetheless general ones about the value of swimming upstream in order to observe the forgotten odysseys of crates, to witness the secret life of warehouses and hence to mitigate the deadening, uniquely modern sense of dislocation between the things we so heedlessly consume in the run of our daily lives and their unknown origins and creators. (ibid.)[11]

Notably, both de Botton's story and Marriott and Minio-Paluello's tale narrate the emergence of the object/animal-as-food in chronological order, reinforcing the temporal trajectory moving *toward* a specific goal, even as the trajectory

---

[11] De Botton's essay can be read alongside Elspeth Probyn's 'Swimming with Tuna' to think through globalization and its repercussions for everyday practices, for example, eating fish (for those who eat fish). I have taught these texts in conjunction in various contexts to think with food, animals or broader globalization concerns.

itself is reversed. Which is to say, rather than radiating out from the object at the centre of their respective stories, going backwards in time, they both prefer to jump back to an imagined beginning and proceed chronologically from there.

The 'destructive after-lives' referenced by Marriott and Minio-Paluello foreground the trajectories of specific objects *after* their use, or on becoming waste (cf. e.g. Hawkins 2013, Thompson 2013). Such stories and provenance studies alike trace the origins and trajectories of particular objects, that is, they have an intrinsic predilection for specific, discrete objects (or collections thereof). The fascination with such studies and reports suggests an (increasing) awareness for 'matter matters'. Whilst they are effective in evoking a sense of the various stories and livelihoods that feed into the object at hand, the set-up of the stories is such that it has a preordained goal to which all paths lead, and, therefore, an established temporal and spatial patterning. That is to say, the tracing backwards (or forwards) of an object in such stories *requires* a step-by-step account through stages and locations in the production or procuration of a specific object, and all such steps *must*, by virtue of this kind of story, lead inevitably from the past to (or away from the present to an imagined future of) the specific object at hand. Again, the connections between spaces and times, between livelihoods and other matters, are thus rendered in a uni-directional, predetermined trajectory, when the emphasis is placed on objects (rather than materials). This gives rise to a clear narrative structure – even the quasi-detective story, as in Hohn's *Moby-Duck* (see also Chapter 2) – but stops short of eliciting the messiness of the entanglements of plastic.

Emergent stories, convergent stories

Massey's 'simultaneity of stories-so-far', developed at length in her monograph *For Space*, suggests a reading of interconnectivity that reaches through and beyond such uni-directional, predetermined trajectories in order to give account of the heterogeneity of space and stories that work against (otherwise often monolithic) 'straightforward' narratives. A politicized sense of the interconnections that form the spaces of our lives is predicated in her work on the following three propositions: Firstly, 'that we recognise space as the product of interrelations; as constituted through interactions, from the immensity of the global to the intimately tiny' (Massey 2006: 9). Secondly, 'that we understand space as the sphere of possibility of the existence of multiplicity in the sense of contemporaneous plurality; as the sphere in which distinct trajectories coexist;

as the sphere therefore of coexisting heterogeneity' (ibid.). And, thirdly, 'that we recognise space as always under construction' (ibid.).

The uni-directional linearity of the trajectories traced in Botton's story and in Marriot and Minio-Paluello's account (which, in the context of this chapter, is more pertinent) fall somewhat short of depicting the density and simultaneity of spaces and stories Massey's theoretical framework privileges. Given the capacity of materials to form relations across spatial and temporal dimensions, as I argued in the introduction, it is somewhat contra-productive to isolate particular trajectories of narrative accounts of plastic in linear patterns. Plastic is material, beyond object(ion). Plastic interrogates and informs as a 'simultaneity of stories-so-far' and as a compression, or coagulation, of practices and imaginations.

One way such a shift beyond a specific object is imagined in cultural forms is by stressing both the singularity and multiplicity of the moment or event. This is to say, the specificity of an encounter with plastic is elicited at the same time as its 'universality', its placedness as well as its potential 'universality', its uniqueness as well as its ordinariness. Two written accounts, one more academic ('Reflections of an Unrepentant Plastiphobe: An Essay on Plasticity and the STS Life' by Jody A. Roberts, from the same volume as Marriott and Minio-Paluello earlier) and the other more literary (*How to Get Filthy Rich in Rising Asia* by Mohsin Hamid), negotiate these impulses through depictions of illness. Both accounts develop interconnectivity in order to elicit a comprehension of plastic relations beyond uni-directional trajectories and binary logics of approval/dismissal or good (plastic)/bad (plastic).

The use of plastic as a material of sterility – keeping infection and contagion at bay – in the medical environments is described in these two accounts alongside less 'desirable' qualities of plastic (the tendency of plastic to become waste after a short lifespan as a useful object, such as explored earlier with respect to cling-wrap). In between drafting this chapter and the revisions, a pandemic happened: Our recent encounters with airborne pathogens and the various plastic borders we employ to protect ourselves (face masks, PPE) or to register illness (plastic swabs and encased rapid antigen tests became a daily event in my household), as well as the waste generated thusly make our complicated relations with plastic evident. This necessitates an account of plastic that takes its specific entanglements with modernization and modernity into account.

As noted earlier in this book, one aspect of modernity that makes it difficult to define is that it appears to continue on (we remain within it) even as it points away from itself for its referent point (to the past, from which we are distancing ourselves): we are, in modernity, continually slipping away from an imagined origin.¹² This conceptualization maps directly onto plastic (and concrete) as modern materials – their invention and uses marks a shift in our relations to the planet and their capacity to puncture time as 'future artefacts'.

Another, related, problem is the way in which modernity presupposes a *particular* trajectory and the way in which this patterning of modernity is mapped on the globe (see the section 'Modern materials' in 'Introduction'). Thinking with alternative modernities looks beyond the monolithic qualities with which Western (or Eurocentric/dominant) modernity is often attributed in order to suggest that there are plural ways of being modern (cf. e.g. Gaonkar 1999, Bhambra 2007). Further, thinking with alternative modernities necessitates and gives rise to a comprehension of material practices that defies a simple categorization into 'good' and 'bad': such a categorization implies a position of authority from which such a judgement can conclusively be made, a universalist (often if not always Western) perspective that is somehow 'outside' of the relations that produce and buttress that very authority. The idea of materiality that pervades this book suggests that such a position is an illusion.

Plastics might be scorned as polluting, as a poisonous practice, but to suggest that they should be banished entirely from the globe entails two, equally problematic and yet diametrically opposed, measures. On the one hand, to suggest 'other places' (read: less 'developed') abandon such polluting practices means that the instance (read: (a manifestation of) the 'West') that offers this suggestion 'knows better'.¹³ It is thus a keenly patronizing gesture. This suggestion also erases the polluting history of the 'West', in particular, the specific entanglements that give rise to the 'knowledge' of plastic's polluting capacities and to concomitant, possibly less negatively connoted, knowledges of hygiene (amongst others). On the other hand, it also suggests – and this insight does not rest comfortably with the first – that there is a very specific goal of modernization/modernity that is desirable, that this goal is universally desirable and that it is possible to take shortcuts in order to reach this particular goal: a goal, again, defined by the 'West'. The knowledges accrued in the 'old' trajectory, this assumption implies, can be acquired without the very practices that gave rise to these knowledges.

---

[12] The 'we' here is as a contentious construct. See note in 'Introduction' and elsewhere. Here, I'd like to bring in Raj Patel and Jason W. Moore's formulation of 'the cumulative actions of everyone able to read this sentence' (Patel and Moore 2020: 207) to approximate the 'we', or perhaps more to the point: everyone who is likely reading this footnote.

[13] See also Young (2003: 2) and the discussion in Chapter 3.

Considering specific practices (here, material practices involving plastic) not as signs of a modernity monolithically understood, but rather as one-of-several possible paths of modernization necessitates and gives rise to *multiple trajectories* through space and time. The authors I discuss – Jody A. Roberts and Mohsin Hamid – use descriptions of multiple connections and nodes to evoke what might be considered an ecology, or mesh, or network, of plastic. Or, as becomes more pertinent in the analysis of Hamid's novel set in 'rising Asia', and thus foregrounding several of the issues giving rise to the plural of modernity, and to use Massey's turn of phrase: a simultaneity of materialities-so-far. Condensed materials, complex relations.

The plastiphobe's dilemma (signs taken as wonders i)

I turn now to Jody A. Roberts's contribution to the volume *Accumulation*, entitled 'Reflections of an Unrepentant Plastiphobe: An Essay on Plasticity and the STS Life'. This chapter is written in a self-reflective, almost ficto-critical, mode. Roberts's attention to endocrine disrupting chemicals (EDCs) is symptomatic of the multifarious intrusions and insertions of plastic into the everyday lives of (US) citizens of modernity and is the basis of his self-ascription as a 'plastiphobe'. In the chapter, he outlines the extent to which he (and by extension, his family) take care to dramatically reduce the presence of plastic in their lives, particularly during his wife's pregnancy. This, he explains, is in order to reduce exposure to the 'unruly technologies' (Roberts 2013: 124) of plastic and the effects on their (as yet unborn) child.

Upon birth, Helena, his daughter, stops breathing and is attached to an apparatus of plastic tubing: 'She was covered in soft, flexible medical tubing – the kind infamously full of phthalates. Nearly everything that came into contact with her passed through this tubing' (ibid.: 127). Feeding tubes, plastic bladder bags, plastic walls of the bed – plastic permeated her early life, first in the NICU (Neonatal Intensive Care Unit), later in their home. Roberts notes that the plastic presence was not just for/of Helena, but also instrumental in the sustenance for the new parents (e.g. food wrapped in plastic or stored in plastic containers). He realizes that 'the plastics that populate my everyday life and that fill me with such anxiety also help to make Helena's life possible, but we resist the simple dichotomies imposed on us. The plastics are not simply life saving or a threat: they are both' (ibid.: 130). The dualism of good/bad (where plastic was clearly 'bad') collapses in this medical modernity, much like the dualism of body/environment in Smith and Lourie's image of the body as sponge in *Slow Death by Rubber Duck* (cf. Smith and Lourie 2009: 2, see also Chapter 2).

For Roberts, knowledge becomes 'a sign of the shifting relationships of my life and world' (Roberts 2013: 130). It becomes 'plastic': malleable, mouldable, compliant, flexible, pliant and so on. This notion of 'shifting relationships' becomes most apparent in the following passage, quoted at length:

> It wasn't just the plastic (potentially) accumulating in Helena's body that bothered me; it was all of the plastic flowing into the world – accumulating in large heaps in the trash cans that were emptied several times a day. These mounds of plastic that each day and night filled the trash cans … left me feeling strangely deflated and defeated. … Plastic syringes, bladder bags, pre-measured and mixed formula, tubing and all the other plastic miscellanea all headed to another location where the accumulated matter would likely be incinerated, which itself would lead to the production and accumulation of new compounds in that environment before molecules of dioxin and other persistent organic pollutants would ride the currents of water and wind to the coldest locations on earth to be deposited, consumed and deposited again in the fat of a seal, polar bear or human being. (ibid.: 128)

In this passage, interconnectivity and its boundlessness are negotiated through the material of plastic. An environmental sensibility is drawn out of the myriad ways in which plastic forges connections – not only between Roberts's daughter and her environment, but also through Roberts's own awareness of the pathways of (post-consumer) plastic reaching out into the world at large. Helena's health is entangled in the dioxins and POPs (persistent organic pollutants) to come: it shows, to reference the title of the volume this essay is included in, an accumulation of plastic waste *yet to come*, a 'future artefact' as mass noun.[14]

Crucially, the scenes evoked by Roberts in his account of his family bear witness to, and carry the weight of (cf. Whitlock 2015: 8), another kind of accumulation of a less obviously tangible kind: the accumulation of practices of medicine. This accumulation is not visualized or rendered explicit in his account. Indeed, in most representations of medical practices, interventions are not accompanied by a tangible or visible history of prior interventions or procedures: The accumulation of practices is present perhaps only to medical historians, whose archival work

---

[14] The accumulation of plastic waste *yet to come* might not 'come' directly to Roberts and his family – 'Opening the trash can, I can't help but think of that strange phenomenon variously known as the Pacific Trash Vortex, Great Pacific Garbage Patch, the Eastern Garbage Patch or simply the Plastic Soup' (Roberts 2013: 130) – that is, there is some distance between Roberts and these locations.

might reveal the depth of knowledges accumulated in specific procedures, or to the avid watcher of historical medical drama (for instance, Cinemax's *The Knick* (2014–15)). This accumulation of practices, medical interventions and trial-and-error is also present in the abandoned face mask, which at the time of revisions, I continue to encounter regularly. Instruments of medical interventions hold the history of shifting and accruing medical interventions within them, a history of the failures and successes of previous interventions and instruments. Material, again, stretching beyond the confines of the object.

Pluralities of plastic (signs taken as wonders ii)

Having illness treated in the hospital is an encounter with modernity. It is also evident in the second example I wish to turn to, a scene from Mohsin Hamid's *How to Get Filthy Rich in Rising Asia*. Hamid's novel is styled as a self-help manual and is set, as the title suggests, somewhere in Asia.[15] The scene in question is placed towards the middle of chapter ten ('Dance with Debt'), where the unnamed, male protagonist has a second heart attack whilst in hospital. Here, as in Roberts, the individual is revealed as dependent on material relations through the event of critical illness, where here it is the protagonist's prior generous donations to the hospital that facilitates his treatment. The following passage narrates the cyborg nature of the intensive care unit, blurring the boundaries between human and machine:

> You are fortunate that your second heart attack takes place in the ICU. When you regain consciousness, you have become a kind of cyborg, part man, part machine. Electrodes connect your chest to a beeping computer terminal mounted on a rack, and a pair of transparent tubes channel oxygen from a nearby metal tank

---

[15] Some critics, for instance, Theo Tait (2013), are very quick to identify 'Asia' as Pakistan and the specific city as Lahore, using Hamid's biography to support their case. Whilst the gesture to avoid the 'flattening' of 'Asia' as a homogenous sphere is laudable, the specifics of this novel – and, more recently, the first setting of Hamid's *Exit West* (2017) – suggest that it is this *unspecificity* of place that drives this narrative as allegory or indeed as self-help book. Michiko Kakutani is more circumspect, noting it is 'set in an unnamed country that resembles Pakistan' (Kakutani 2013), and later dubbing it 'not-Pakistan-exactly' (ibid.), although the phrases do of course still establish the link even as they deny it (pink elephants!). Sharae Deckard, in contrast, suggests, the 'indeterminate setting might be read as a deliberate attempt to circumvent postcolonial readings of the novel in order to enable more world-systemic comparisons of the dynamics of liberalization, corruption, financialization, urbanization and authoritarian state power in "emergent" economies throughout the Global South' (Deckard 2015: 244). Deckard's mobilization of world systems critiques the uni-directionality of postcolonial, through which the various sites of the 'postcolonial' can only be accessed via a centre (the colonizers).

to your nostrils and fluids from a plastic pouch into your bloodstream through a needle taped at your wrist. You panic and start to flail, but your limbs barely move and you are gently restrained. ... You understand, though, that for the moment this apparatus and you are inseparable. (Hamid 2013: 185)

The discrete, individual-izable, identity of the protagonist is questioned in this passage. The passage also explicitly mentions two materials (metal, plastic), foregrounding the inseparability of human life from the materials that surround it in this state of emergency (in the emergency department). Continuing life, here, is shown to be dependent on a number of interventions into the body of the protagonist.

The passage then continues to elicit the 'experience [of] the shock of an unseen network suddenly made physical':

The inanimate strands that cling to your precariously still-animate form themselves connect to other strands, to the hospital's power system, its backup generator, its information technology infrastructure, the unit that produces oxygen, the people who refill and circulate the tanks, the department that replenishes medications, the trucks that deliver them, the factories at which they are manufactured, the mines where requisite raw materials emerge, and on and on, from your body, into your room, across the building, and out the doors to the world beyond, mirroring in stark exterior reality preexisting and mercifully unconsidered systems within, the veins and nerves and sinews and lymph nodes without which there is no you. (ibid.: 185–6)

His stasis – the 'gently restrained' body – and his status – alive but critically ill – are dependent on a network of carefully choreographed movements, human and material alike. Here, it is through the evocation of various forms of logistics and infrastructure (informational and material) that a sense of interconnectivity becomes evident. The protagonist, 'precariously still-animate', is simultaneously shown to be the centre of these (logistical, infrastructural) movements, at the same time as being entirely peripheral and external to these movements, as well as, of himself, a hub of activity (through the various pathways of blood, neural signals, hormones): body meets world as body parallels world.

In contradistinction to the passage from the Roberts chapter earlier, which stresses the consequences of the overabundance of plastic in terms of waste, Hamid's passage relies much more strongly on the wonders of modernity rather than the negative (environmental) effects. Plastic is not especially foregrounded in this passage and yet it is peculiarly omnipresent throughout the novel. In the passage cited, plastic is the material of some of the medical tools (and, crucially, one of the two materials that are usually used for the constructions of the stent that a surgeon will later place in the protagonist's heart). Earlier in the novel, plastic figures as the material of the DVDs the protagonist delivers throughout the unnamed city, including to the 'pretty girl' (who will become the 'love of his life'). It is also, crucially, the material of the bottles which, filled with water, fuel his career and success (on the way to getting 'filthy rich in rising Asia'). In this novel, it is through the various manifestations of plastic that a sense of (alternative) modernity is negotiated, giving rise to entanglements with this and other materials that cannot be reduced to simple binaries (good/bad, progressive/regressive, etc.).

Packing and unpacking plastic

'Unpacking' the title of Hamid's novel is a first step to elicit the various manifestations and attributions of plastic in his novel. The tension articulated in the title – *Filthy Rich* – emerges as a material concern, most obviously articulated in the form of plastic bottles, which themselves play on the fears of filth, dirt and unsanitary water. The title also figures the conceit of the self-help guide: In Hamid's text, this emerges as an account that utilizes the indeterminateness of the form of second-person address to stress the complicity of the reader in modernizing tendencies and material practices.

In her analysis of Hamid's novel, Angelia Poon suggests that the 'object of the novel's satire is the capitalist, neoliberal notion of the self that is predicated on an over-weening sense of control and ultimate agency' with 'material affluence as the goal' (Poon 2015: 140). The idea of 'material affluence' (and its near homonym 'effluence') is productive: this is an *excess*, a flowing (or overflowing), of materials. Poon notes the generic traditions of the self-help text in (Victorian) desires for 'upward mobility'. The concomitant belief in, and whole-hearted subscription to, trajectories of development understood in uni-directional, linear steps towards a 'better state of being', are also symptomatic for the larger projects of modernity and development, critiqued through alternative modernities. The genre of 'self-help' is an apt location for satirizing the unsavoury effects of uneven development (where the self becomes a site for the accumulation of materials and/or wealth): Its 'effluence' or 'effluent' is foregrounded, in this particular case, by the *Filthy Rich* of the title.

To turn to the filth. When the reader first encounters the protagonist, he is suffering from hepatitis E: 'Its typical mode of transmission is fecal-oral. Yum. It kills only about one in fifty, so you're likely to recover. But right now you feel like you're going to die' (Hamid 2013: 4). The link between affluence and effluent, forged through water, becomes once more explicit at the beginning of chapter five ('Learn From a Master'): 'Where moneymaking is concerned, nothing compresses the time frame needed to leap from my-shit-just-sits-there-until-it-rains poverty to which-of-my-toilets-shall-I-use affluence like an apprenticeship with someone who already has the angles all figured out' (ibid.: 78). The shift away from this particular kind of flow, and the projects of sanitization and hygiene, is linked to the larger project of modernization throughout the novel (including the scene in the hospital, as mentioned earlier).

Throughout *How to Get Filthy Rich in Rising Asia*, plastic acts as a conduit of modernity. Present at numerous conjunctions of materiality and modernity throughout the novel, getting 'filthy rich' with plastic (ultimately in the form of plastic bottles) is foregrounded, or rather, materially pre-empted, by further practices and objects in the novel. The pirated DVDs the 'you' delivers as one of his first jobs, for instance, are made of plastic: not only does the 'you' become thus familiar with the logistics of delivery, pre-empting his later 'empire' of bottled water, the DVDs crucially function as plastic bearers of a (cultural) modernity. Another precursor to the bottled water, this time revealing the propensity of consumers to purchase packaging as much as the goods themselves, is the protagonist's role as 'a non-expired-label expired-goods salesman' (ibid.: 99). Like his work as a DVD delivery boy, this job stresses the importance of the logistical network and references (lack of) infrastructures as material grounds that can be monetized. It also picks up on the trope of water and health and thus effluence and, ultimately, affluence. Crucially, access to clean water is figured as a function of capital, financial as much as cultural. As in Chapter 3, water emerges as an actor in modernization projects, manifesting particularly in infrastructure. The concrete consequences explored in the previous chapter become plasticized here: as bottled water.

The stratification of water relies on its packaging, both the material (the bottle, the labelling) and its branding. Indeed, Hamid's unnamed entrepreneur fills boiled water into plastic bottles salvaged from restaurants in the first stages of his bottled water empire. When buying bottled water, 'what the consumer is really buying *is* a package' (Hawkins 2011: 536), Gay Hawkins argues in 'Packaging Water: Plastic Bottles as Market and Public Devices'. For the consumers of the early stages of Hamid's protagonists' foray into 'plastic water', this is certainly true. The container is not so much of interest as an object, but as a node in the flows of materiality – both plastic and water – and as a practice of modernity.

This is true for 'modernized' parts of the world, given the ready availability of clean water to most consumers of bottled water, but maps differently in the 'developing world' or as the title has it, 'Rising Asia'.

In *Plastic Water*, Gay Hawkins, Emily Potter and Kane Race examine a number of sites throughout Asia as case studies for their analysis of bottled water. Of the three sites and case studies examined in detail by Hawkins, Potter and Race, it is Chennai (in the central south of India) that resonates most closely with the novel. In the chapter entitled 'Enacting Water Scarcity in Chennai', they observe that 'this urban water context thus reframes the bottle from a convenience or leisure consumption item into a significant participant in potable water provision' (Hawkins, Potter and Kane 2015: 78), making some binaries, like private and public, or other ways of thinking about bottled water in the West inadequate, 'incapable of capturing the mess and fragmentation of water supplies or the ways in which multiple modes of water provision coexist and interact' (ibid.).

In *How to Get Filthy Rich in Rising Asia*, the bottled water infrastructure only starts to give way to reliable (potential) potable water from the tap in chapter nine ('Patronize the Artists of War'), a chapter which itself is cut off by the event of the heart attack described earlier. The protagonist and his soon-to-be-estranged brother-in-law start off talking to a member of the military complex about providing potable water from the tap. The brigadier notes the dreams and aspirations of this project as follows: 'Drinkable water. It'll be like you've entered another country. Another continent. Like you've gone to Europe. Or North America' (Hamid 2013: 164). This 'elsewhere' compounds into an 'elsewhen', where potential trajectories of emplaced development point to other places. This desire for (westernized) modernity is not achieved (by the protagonist) in this novel, but it nevertheless figures as a desirable outcome. And, furthermore, its configuration in these terms implicates the (implied) reader in the tale, who, presumably, lives in these locations and enjoys the benefits of such plastic practices.

Crucially, Hamid employs a relatively uncommon narrative situation for his tale, namely the second person pronoun 'you'. In English, this pronoun is both singular and plural and both familiar and formal. Within a narrative situation, particularly sustained over the length of a novel, the use of the 'you' is rather strange. Brian Richardson argues that 'second person narration is still too rare, too unusual, and too discordant to be conventionalized or domesticated; it still has the power to produce a bracing sense of estrangement as standard distinctions between narrator, character, narratee, and actual reader are conflated' (Richardson 2013/2014: 53). Hamid's use of the second person goes against expectations for a novel-length text, and, at the same time, 'more easily accommodates positions of identification for readers because it always carries a residue of its appellative and conative functions from real-life usage', as Jarmila Mildorf suggests (Mildorf 2016: 109).

The you's 'slippery referentiality' makes it adaptive, flexible, pliant, indeed *plastic*. In Hamid's novel, it might refer to the 'unknown protagonist', to the reader, and be used in a sense also covered in English by 'one' (so an impersonal, non-deictic use). In the framework of a (satirical) self-help book, the 'you' most obviously addresses the reader; however, slips into aphorisms and platitudes work to conflate the reader position with a generic 'you' (i.e. the non-deictic use aligned with 'one'). And, obviously, it refers to the protagonist.[16]

In the opening pages, Hamid mobilizes both possible referents for the address 'you': Initially leaving the identity of the 'you' open (to suggest both reader and protagonist) means that the affect of the boy's 'anguish' can be used to great effect, even as the two referents uncouple, I'd suggest, between the two following sentences:

> Your anguish is the anguish of a boy whose chocolate has been thrown away, whose remote controls are out of batteries, whose scooter is busted, whose new sneakers have been stolen. This is all the more remarkable since you've never in your life seen any of these things. (Hamid 2013: 4)

What is striking about this concatenation of seemingly everyday goods/toys is the way that it establishes 'belonging' to the (singular) project of modernity and development through material goods predominantly made of plastic: remote controls are (partly) plastic, sneakers are (probably) plastic, chocolate often comes in plastic wrappings, and toy scooters have at least some components made of plastic.

The radical denial of such identification, the uncoupling of the dual referents of the 'you', reinforces the complicity of the reader of the novel with the processes of production and consumption, with the materials of modernity, and with the unevenness of access to such goods. As Heather Davis asserts, 'There is no way to extract one's life in the twentieth century from plastic. This is true for people across economic classes and geographies, even if the objects we interact with and the ways we do it remain stratified' (Davis 2015: 349). Insofar as the reader acknowledges the address of Hamid's 'you', then, the reader also might find

---

[16] The protagonist, due to many identificatory factors – including gender, class, nationality (even if this is left vague) – does not map onto me, that is, Kylie, the writer of these words, nor (perhaps) you, the reader of these words, either.

themselves querying their implication in networks and flows of the materials of modernity.

Coda

Plastic materialities manifest agency in temporal, spatial, affectual and narrative contexts. They shape the world as they are shaped by it. Plastic packaging fulfils several functions of material and semiotic consequence. These 'consequences' are not separable from one another: Like cling-wrap, they have a tendency to cling to all sorts of matters.

Plastic packaging defines objects, by distinguishing what is contained from what is not contained. As addressed in detail in conjunction with the artworks of Christo and Jeanne-Claude, packaging comprises the border at the same time as marking it as such (it is simultaneously object, border and practice, if you like). The packaging, seemingly backgrounded, becomes a constitutive component of perceiving and interacting with object, environment, and the junctures between the two.

In the case of biological matter, plastic packaging often functions to preserve, but at the same time, plastic preserves itself. Again, packaging-as-border refuses its backgrounding through its very materiality. In plastic packaging, the duration, or time span, of the material preserved and the preserving material are vastly different, the latter outlasting its 'functionality' with respect to the former by orders of several magnitudes (a cling-wrapped sandwich or serving of cooked food, for instance, might last for several days; the cling-wrap itself might last for several decades or centuries, we don't really know, cf. Liboiron 2021: 17, fn 64).

This border function is, however, not absolute. As also explored in Chapter 2, the plastic of the packaging does not possess material 'integrity' of the scales its presence in the environment would suggest. Its varying capacities to interact with biological organisms, for example by mimicking endocrines in humans, show plastic to be an active agent in its surroundings. Despite the negative (environmental) connotations of these arguments, the role of plastic on our planet, in our lives,[17] should not and cannot be entirely negative. Plastic, as shown in conjunction with the second part of the chapter, with Jody A. Roberts's

---

[17] The doubling of 'our planet, our lives' is deliberate, to suggest a slip between the humans that inhabit the planet and the human lives that are covered by my analysis. These are not identical 'ours', and, again, my own interpolation will not necessarily extend to all my readers equally.

essay and Mohsin Hamid's novel, is a crucial material for thinking through (alternative) modernities and (life-saving) practices of modernization. Plastic's capacity to preserve and also to act as a border to biotic matter – the qualities that make it so slow to biodegrade – are also the very qualities that make it attractive for medical practices of modernity. Plastic emerges as a particularly *modern* material: not only signifying modernity, but also constituting it. The 'invention', or 'coming-into', of plastic, examined in this chapter through a particular manifestation – packaging – emerges, finally, as relations of quite significant (spatial and temporal) expanse.

5

Concrete ruins

Introduction: Concrete ruins

Concrete ruins. The two interpretations of this phrase push back on each other: As a nominal phrase – ruins as a noun and concrete as a modifier – the phrase stands alone, waiting for something to happen; and, as a most simple sentence – concrete as a noun, ruins as a verb – it is a process already in action. Both interpretations come into play in the following chapter, drawing out the relations of materiality as ruins and as agents of ruining, manifesting in specific dimensions and sites. These encompass factories, high-rises, sarcophagi and bunkers; structures for living, for remembering and for containing. The locations of this chapter are various, encompassing Haiti, the UK, the United States, Bangladesh, Ukraine, Finland and the Marshall Islands.

The idea of 'ruin' might bring to mind a set of images, conglomerating around sites of the real world (such as Angkor Wat, Macchu Pichu or the Acropolis) and of the imagination (in the Western tradition, the list might include Percy Bysshe Shelley's 'Ozymandias', Atlantis, the Tower of Babel). Such 'prototypical' ruins evoke a romantic sense and a Romantic tradition, which are productively thought together, and are often imagined as, to borrow the words of Laura Ann Stoler, 'enchanted, desolate spaces, large-scale monumental structures abandoned and grown over' (Stoler 2008: 194). And yet, as Stoler goes on to caution, we need to be wary of an all-too-nostalgic stance, to be sure not to gloss over 'structures of vulnerability' (ibid.) that are sustained by (imperial) formations.

Ruin, as debris or as accumulating masses, has a political dimension. This extends from – but is not limited to – questions of how to 'deal with it', especially in those circumstances where we might speak of ruins as waste. As Tim Edensor argues,

> One of the characteristics of power is the ability to make decisions about what is required, and therefore what objects get to be produced and in what form. It therefore also becomes clear that one of the lineaments of power is the authority to make waste, to decide what is no longer of use and disseminate common-sense ideas about what ought to be over and done with. (Edensor 2005b: 105)

This political dimension of ruins manifests as a set of logistical considerations with material effects: Asking, for instance, where and by what means and through what kind of payments to whom ruins are removed from the quotidian realities of one (set of) people to another (set of) people is not just a tracing of the shifts of the materials. It is also a question of value, of value extended to the materials, of value extended to the people who are 'left with' the materials and of value extended to practices that engage with ruins, including those which must 'make do' with what is 'left over'. We might think of some people in some countries exporting their plastic waste to other people (in other countries), or of ship breaking yards, as in India, Pakistan and Bangladesh (amongst other sites), as way in which these considerations become logistics of materials.

Ann Laura Stoler, in her introduction to *Imperial Debris*, offers a succinct statement of this potential when she proposes considering '"ruination" as an active, ongoing process that allocates imperial debris differentially and *ruin* as a violent verb that unites apparently disparate moments, places, and objects' (Stoler 2013: 7, emphasis in original). For my book, with its concern for the materiality of materials, ruin's relationality is preferable to the accumulation of discrete forms suggested by debris (in Stoler's title). Ruins are, Stoler writes elsewhere, the 'corroded hollows of landscapes, ... the gutted infrastructures of segregated cityscapes', they are not 'inert remains but ... vital refiguration[s]' (Stoler 2008: 194). The agential capacity of concrete ruins, of concrete to ruin, forms the central concern of this chapter.

Stoler's proposal is to consider ruins as not what is 'leftover' but what people are 'left with' (ibid.). Ruin might also be a site *within* the body, as in the case of the 'imperial ruins' of the Vietnam war and the widespread and indiscriminate use of chemical herbicides (known as 'Agent Orange', cf. e.g. Stoler 2013: 25–6).[1] A shift in agential capacities of human and non-humans alike slowly becomes evident. Stoler expands on this in the introduction to *Imperial Debris*, by suggesting that 'we might turn to ruins as epicenters of renewed collective claims, as history in

---

[1] This entanglement of toxicity and ruins will be revisited briefly later, when I turn to sarcophagi and other practices of (ostensibly) containing nuclear waste. I consider precarity and toxic waste in post-industrial ruins in more detail in 'Displacements' (Crane 2021b).

a spirited voice, as sites that animate both despair and new possibilities, bids for entitlement, and unexpected collaborative political projects' (ibid.: 14). Ruins, then, are products of the past, selectively permeate the present, and shape an uncertain future. They are obvious examples for temporal shifts. Accordingly, ruin in this chapter draws on these deliberations of the processual character of the word and on the ideas evoked by the term 'ruination'.

In his book *Concrete and Culture*, Adrian Forty argues that 'concrete can be more accurately described as a *process* than as a *material*' (Forty 2012: 44, emphases in original). As also noted in the introduction, Forty suggests:

> As a material, reinforced concrete is not dissociable from the works made from it – one cannot take a piece of reinforced concrete, a 'sample', to show what the structure will be made of, for reinforced concrete only happens when the work is cast and the network of forces between steel and concrete becomes 'live'. (ibid.: 51)

As he further asserts, concrete has a complex temporality, noting that 'the emphasis in all the material originating from the cement and concrete industry [on] the "present possibilities" and "future potential" of concrete'. 'Over and over again', he stresses 'we are told that concrete is a material full of possibilities, whose full potential is yet to be realized' (ibid.: 86). Concrete's temporality, in Forty's reading, is always one of another time – 'elsewhen'. For him, it is a temporality projected into the future and masking the present: The 'reluctance to treat concrete as a gradually evolving practice is particularly evident in how few built works of architecture ever make reference to concrete's own history' (ibid.: 87). In line with considerations of the Anthropocene (see Introduction), concrete objects are concrete reminders of our current modernity, potentially to be found by future archaeologists. The tensions and complexities it embodies – independent of how we might consider these today – are open to future interpretations; even as, or perhaps particularly as, they become ruinous.

Moving worlds: Concrete fissures, island interstices

The interplay of concrete structures, in particular the crucial moments in which the flows of ruin coalesce as they fall apart, is crucial to Dany Laferrière's *The World is Moving Around Me*.[2] As the subtitle declares, this is *A Memoir of*

---

[2] The book was originally published as *Tout bouge autour de moi* in 2011; I read the translation by David Homel, published by Arsenal Pulp Press in 2013.

the Haiti Earthquake, which occurred in January 2010. For this chapter, it is Laferrière's suggestion that 'concrete was the killer' (Laferrière 2013: 16), which impels analysis: here is a material afforded a malignant – murderous – agency.[3] Elsewhere, Mark Miodownik suggests that 'the extent of the devastation due to the 2010 earthquake in Haiti was blamed on shoddy construction and poor-quality concrete' (Miodownik 2014: 69). Concrete's agency emerges in Laferrière's work through direct attributions and through the structure of the memoir as a whole.

Laferrière's memoir comprises a series of short vignettes. The short vignettes, many less than a page long, include titles such as Projectiles (from which the 'concrete was the killer' quote is taken), Silence, Objects, Time, Place, Forty-Three Tremors, The Concrete Trap, Taking Stock, No Place, The Energy of Things, Malaria, A Body Quake, Wood, New Landmarks, Electricity, New Art Forms and The Tenderness of the World. Such titles already suggest a myriad of agencies afforded to materials, as well as the tensions and stresses of place-making through representational practices, such as the writing itself.

This composition of the text as a collection of vignettes gives the memoir a fragmentary structure, as a collection of rubble, that is, as parts of larger stories or as various attempts to approximate a story. These fragments are contained within the scope of the book but continually push back on this containment. The fragmentary structure reckons with the various approaches, or to use Sara Ahmed's phrase, angles (cf. Ahmed 2010), to the material at hand.[4] It does not suggest a 'de-fragmented' whole in any way that can be reassembled: The many

---

[3] From the fragment 'The Concrete Trap', also early in the book:

> A lady who lives nearby spent all night talking to her family still trapped beneath a ton of concrete. First her husband stopped responding. Then one of their three children. Later, another. ... More than a dozen hours later, people were finally able to rescue the baby, who had been crying the whole time. When he got out, he broke into a wide smile. (Laferrière 2013: 29–30)

[4] In 'Happy Objects', Ahmed writes: 'So we may walk into the room and "feel the atmosphere" but what we may feel depends on the angle of our arrival. Or we might say that the atmosphere is already angled; it is always felt from a specific point' (Ahmed 2010: 37). In this essay, Ahmed is exploring affect as something 'sticky'; the notion of angles is subordinate to the way that moods 'affect' approaches. I find it a valuable way for thinking through the materiality of encounters, particularly for the way it is suggestive of spatial dimensions.

Consider, for instance, that ultra-black paints, such as Vantablack 2.0 or Stuart Semple's Black 2.0–4.0, work to mask the contours of the objects to which the paint is applied by absorbing more light rays than other black (or otherwise coloured) paints: the visual impression of the object is a direct result of the angles of light being refracted differently. Perception – here, visual perception – is a consequence of angles. Karen Barad's explication of apparatus entanglement in physics experiments in order to articulate their concept of intra-action similarly takes account of approach, although in a different manner to the (predominantly aesthetic) point I wish to make here.

surfaces offered by the fragments far exceed any singular entity, with more of the material exposed than if it were whole, a crumbling, if you like.

A further interpretation of the structure, one which I believe is to be layered with the above-mentioned interpretation of it as ruins or rubble, is as a hesitation, a stepping back from a claim of completeness or cohesion of the story. This gesture allows different stories (by different authors) to coexist. Laferrière is in fact careful to stress the limits of perspective and the ethics of story-telling. On watching a student from Miami take photographs (like a lot of people who 'come to help' but also 'try to capture the suffering on film', Laferrière 2013: 116), Laferrière notes, 'I sat down on a low wall to write. How to describe a scene like that? He took just one picture' (ibid.: 117). The sentiment here is one of a reluctance to contain the story within simple or small structures, a reluctance that Laferrière reiterates through various vignettes in recollected conversations with his nephew.

As the blurb declares, Laferrière's nephew (also called Dany) asked him to not write the book. His nephew's rationale is that whilst the elder's generational event was the dictatorship, his own generation has a kind of ownership of the events of the earthquake (cf. ibid.: 50). Whilst, obviously, the elder Dany does not abide by this request (as the book itself attests), the recollection of this exchange foregrounds ethics of (positions of) speaking and of the ownership of stories. Who gets to tell what kind of story, how and to whom? Such questions, in turn, give rise to awareness of the multivalences of stories, the different ways of approaching and telling them – Doreen Massey's 'simulteneity of stories-so-far' (see Massey 2006, as well as my introduction and Chapter 3). This line of questioning finds a manifestation, again, in the vignettes. The rubble, the debris, the broken lives and perspectives find their objective correlative collated and presented in this broken form.

There are many instances within the vignettes themselves that address concrete, and, more broadly, materiality and the institutions which concrete has come to represent. It is through these *concrete* evocations of debris that a sense of a structural *ruin* emerges. Take, for instance, the concrete structures of Laferrière's Haiti, which are associated with projects of modernization, that is, specific institutions: 'We gazed with wonder as the disaster revealed a nation whose rotten institutions prevent it from coming into its own. When those institutions disappeared from the landscape, even for a moment, we discovered a proud yet modest people through the clouds of dust' (Laferrière 2013: 27).

The conceptual link between the concrete manifestations of institutions and the organizational entities they represent – a metonymical relation – is blurred through its differing materialities: Institutions are rotten (a biological qualifier, with a precursor in Shakespeare's *Hamlet* (2001: I.4)), are a fixture in the landscape and can be reduced to 'clouds of dust'. At the same time, institutions are an abstract manifestation of modernization and modernity: A building only becomes an institution through specific uses by specific people.[5] In this passage, Laferrière evokes the Haitian people as a function of a discursively formed nation that breaches its institutions, that is, that exists separately from them. The ostensibly stable link between material forms (institutions as buildings) and their abstract use (institutions as functions) is jolted through the event of the earthquake.

A similar sentiment emerges in another vignette, much later in the memoir, when Laferrière writes:

> The veneer of civilization that I'd been inculcated with went up in smoke – a cloud of dust like the ruins of the city. All that took ten seconds. Is that the true weight of civilization? During those ten seconds I was a tree, a rock, a cloud, or the earthquake itself. One thing was for sure: I wasn't the product of a culture anymore. I had the definite impression of being part of the cosmos. (Laferrière 2013: 86)

Here, civilization (read here as a near-synonym for modernization) becomes equated with the concrete manifestations of a (Western-styled) development. Laferrière, however, does not render this as a moment of existential angst, as a threat to his individuality, but rather as a reminder of the materiality of his self and the connections it forges beyond the bounds of culture or nation. He celebrates the resilience and adaptability of people and organic life, which are developed as tropes in opposition to institutions and inorganic materials: 'The earthquake attacked what was hard, solid, what could resist it. The concrete fell. The flowers survived' (ibid.: 22). In such passages, a certain dualistic relation emerges: people of Haiti/institutions of Haiti; flowers/concrete. Concrete is aligned with non-life, even with the destruction of life. Whilst not equal to the forces of the earthquake, here concrete becomes a sparring partner for the forces of seismological shift.

This oppositionality is not absolute. The fragmentary structure of the memoir, as explored earlier, resists such an absolute oppositionality, which would prove difficult to maintain across the various vignettes. Instead, such tensions are

---

[5] See Brian Larkin's explications of 'The Politics and Poetics of Infrastructure' and in particular for how he articulates 'the postcolonial state's imaginative investment in technology' (Larkin 2013: 333).

multifaceted, full of angles and resist clear compartmentalization. Accordingly, Laferrière observes on the relationality of materials and the positions materials take in relation to the environment:

> Haiti was on the brink of an ecological disaster. No trees, so nothing to hold down the arable land when the hard rains fell. … If there are any trees left standing, it's because concrete has become the favorite building material. But since it failed the earthquake test, the latest talk is of going back to wood, since it's more flexible and resisted the tremors better than cement. That's true enough, but if we go back to wood, we'll risk ecological catastrophe. (ibid.: 113–14)

The complexities of material entanglements are only hinted at here, as are the inherent contradictions of development. Concrete's malevolence, read above as a quality of its agency, is un-dis-entangle-able from, and subject to, greater environmental forces. It is neither 'precondition' nor 'consequence', but rather emergent *with* its materiality. In this vein, the somewhat longer passage from the fourth 'fragment' called 'Projectiles' reads thus: 'A 7.3 magnitude earthquake is not so bad. You still have a chance. Concrete was the killer. The population had joined in an orgy of concrete over the last fifty years' (ibid.: 16). Concrete's agency, I suggest, remains evident, although it must, in the context of the memoir, be read as a function of its use (and abuse) within a specific environment.

A notion of collectivity within a specific environment (the Haitian people) – resisting the containment by narratives of development – is as central to Laferrière's memoir as the structural fragility of the narratives and the *concrete* manifestations through which they emerge. It is thus that the full force ruin is not simply that which is *left* (over), but that which people are *left with*, comes into being. This is rendered in the two key ways: firstly, emergent through shared plights and stories, and secondly, as a sense of memoried accumulation.

In the first passage, Chantal Guy (a Quebecois journalist who has travelled to Haiti at Laferrière's insistence) asks Laferrière: 'What is the value of culture in the face of disaster?' Laferrière notes his response as follows: 'I look around: it's easy to evaluate the situation. The conversations are lively. I hear laughter from time to time. People are looking for some way out. Which makes me think that when everything else collapses, culture remains' (ibid.: 59–60). Laferrière's suggestion that 'culture remains' (ibid.) evokes a surplus, an excess, to material structures. And yet, the 'value of culture in the face of disaster' is the capacity to render 'surplus, excess' from the ruins, present in the vignettes, stories, collected in the volume. These stories are intricately connected to and emerge with the 'landscape of this crumbled city', to which Laferrière notes, 'People have added elements of the old one still present in their memory. For the population whose minds are always in ferment, things accumulate instead of disappearing' (ibid.: 118).[6]

Here, the earthquake does not (just) destroy and lead to the removal of materials (that comprise homes, schools, hospitals, livelihoods). The earthquake, instead, is one event of many that leads to an accumulation, a thickening and 'richening', of life. The *concrete* materialities – both those made of concrete, and the specific materials of life in Haiti – thus coalesce. The memoir nevertheless depicts a concrete world, effected through and affected by ruin.

Factories between ruinous production and productive ruins

As noted in Chapter 3 and in the introduction, concrete has localized and globalized components: The water and aggregate are often locally sourced, whereas the cement (and steel, in the case of reinforced concrete) are products of globalized industrial processes. Similarly, factories manifest industrial capitalism and increasingly globalized flows of finance and materials. As such, they are intrinsically contradictory sites. That is, factories can be understood as material manifestations of the abstract forces that shape most of the world, insofar as most of the world relies on factory-based forms of production. Factories produce (concrete) goods and through this (abstract) surplus; sometimes they produce surplus goods that are on their way to becoming waste. For those who work in the factory, particularly on the factory floor, it is a space of labour, of strain and repetitive manual tasks.

With news reports of factory collapses, such abstractions (or omissions) crumble. The collapse of the Rana Plaza building in greater Dhaka, Bangladesh, in April 2013 is one such (highly medialized) example. The reports of this event substantiate how ruin can be understood in an active sense and forges connections between otherwise seemingly disparate sites, reminding us that the relations that clothe Western bodies extend around the globe. The collapse of the Rana Plaza is at the same time only an event insofar as it is considered as a discrete entity. The factory, as a site of production of surplus value, is located

---

[6] The metaphor of fermentation is particularly powerful. It suggests a process that specifically relies on environmental factors that exceed human control (as in the fermentation of yeast in bread-making, which is always situational, location-specific).

in Bangladesh *as a result of* neoliberal politics and capitalist accumulation. It is a node in a network of the relations that comprise the clothing industry: The clothes 'we' wear are connected not only to the site of sale, but also to a globalized logistics of agriculture/mining (depending on the specific materials, i.e. cotton or wool vs polyester or elastan – plastics), spinning and weaving, cutting and sewing, selling and wearing. The collapse, similarly, is also the result of politics and accumulation, even as the accumulation might be more readily figured as a degeneration, that is, as an accumulation of faults, failures and fissures (or: ruin as verb).

The collapse of the factory, then, is mediated as a discrete event but is indicative of a slower, more accretive violence (cf. also Nixon 2011) – a violence that amalgamates the material and discursive as well as the global and local. Such forms of 'concrete' materiality is figured in a number of ways in Jason Motlagh's contribution to *Virginia Quarterly Review* called 'The Ghosts of Rana Plaza'. The essay revisits the collapse a year after its occurrence and is supplemented by a photo-essay by Atish Saha, including several images depicting displaced slabs of concrete. In the following, I examine the ways in which concrete is depicted in this essay as forming the ruined world.

The Rana Plaza factory is located, Motlagh explains, in '*bastis* – dense neighborhoods of concrete and tin barracks where poor garment-making families live' (Motlagh 2014). Concrete is thus evoked as constitutive of the material reality of the workers of the factory beyond the specific framework of their place of employment. It manifests, at home and at work, as a material that gives form to the structures that shape the daily lives of the workers there, both as workers and as individuals who have lives beyond the constraints of the factory.

Concrete is also present in the factory itself. In Motlagh's description of 'a cascade of calving concrete and machinery', concrete's mass is of proportions that warrant the verb 'calve'. As a verb used to describe the splitting and shedding of large chunks of ice from icebergs and glaciers, as well as giving birth (for large mammals), the image is of a monolithic mass of material cracking (evoked onomatopoetically through the five 'c/k' sounds in 'cascade of calving concrete'). In Motlagh's descriptions of the collapse, the concrete merges with the reinforcing rods to create a 'mound of concrete slabs and twisted rebar' and the 'broken concrete' forms unsettling grounds, requiring stabilization (ibid.). Concrete is evoked as an agent in the event of the collapse, as a 'massive concrete beam that pinned [Paki Begum] down' and later 'muted' Begum's voice (ibid.). Similarly, concrete is also described as preventing another worker, Shahina, from leaving the rubble, first snagging her clothes and ultimately creating a space too tight for her to leave at all. It also works as a material barrier, inhibiting rescue operations. The syntax of such sentences in Motlagh's text affords concrete agency as a subject, an enactor of violence.

In addition to concrete's agency as a massive malevolent force, recollecting the passages in Laferrière's memoir, concrete is also mentioned in its particulate form, as 'cloud of concrete dust'. This evokes a specific kind of image. In particulate form, substances can be inhaled and smelled and thus incorporated into the body. Concrete, as dust, is seemingly the *least concrete* form of concrete. In terms of its prototypical functions, as dust it seems pretty useless: Concrete dust cannot contain or support, structure or withstand. This (lack of) utility is, obviously, an inherently anthropocentric utility. As dust, concrete is neither discrete object nor distinct entity: It *is* its materiality.

The coalescing of material erupts, then, in both the depictions of the Rana Plaza collapse and the Haiti Earthquake, through a violent event. Its agency is no abstract philosophical concept, but a very real, life-threatening, life-taking force in such events. And yet, this agency is present, if dormant, in the structures from the point in time when they are constructed: The event of collapse, like the long duration of ruin-as-slow-violence, are latent in the forming of the structures as a direct consequence of this forming. We have factories producing concrete, concrete forming factories: The entire timeline is imbued with a sense of future ruining as much as future ruins. The relations of slow violence to discrete events like the collapse of Rana Plaza in April 2013 is one that echoes the timelines I articulated with my analysis of cling-wrap in Chapter 4: It is a kind of erupting or collapsing of dynamic and far-reaching relations into a condensed space/time. The events of the collapse of Rana Plaza are sudden (eruptions) of longer processes. This is the kind of folding and intensification of space and time through materiality which I approximate with the term 'future artefact'.

The factories in Dhaka, Bangladesh, are, to a certain extent, functions of the abandonment of the factories in the Global North as capital turned offshore to offset costs of production. I turn, now, to some sites of the Global North to examine the durations of ruin where considerations of the global(izing) factors of the site of the factory are somewhat backgrounded: Tim Edensor's explorations of factories in Great Britain. His attention to the materialities of the abandoned factories warrants analysis in order to more carefully explicate the intricacies of material, decay and ruin.

Edensor is interested in the industrial ruin. The industrial ruin has a temporal dimension that stretches – and draws attention – away from the present: to that which has vanished, but was previously there, and towards an uncertain future (cf. Edensor 2005b: 7).⁷ Its temporality is much slower than the 'event' of the collapse, like the Rana Plaza disaster. The industrial ruin of Edensor's account is located between a 'productive' past and unknown (or 'stagnant') future (where the qualifiers are to be read in terms of capitalist value attribution through production, and not in absolute terms). The temporality of the industrial ruin entails a lack in two directions: a lack of something from the past and a lack of something intended for the future. For Edensor, 'Ruins offer different ways of remembering the past. They are already material allegories of the imperfect way in which the past is remembered, replete with loss and confusion' (ibid.: 170). As Edensor argues, the ruin contests the idea of 'meaningless' space (albeit in a different way to that of Doreen Massey). In his book-length study, *Industrial Ruins: Spaces, Aesthetics and Materiality*, he 'contest[s] the notion that ruins are spaces of waste, that contain nothing, or nothing of value, and that they are saturated with negativity as spaces of danger, delinquency, ugliness and disorder' (ibid.: 7).

Ruins, and the concrete materiality of ruins, emerge as contradictory sites. The presence of ruins is indispensable or indisputable (I want to use the German word *unabdingbar*, for its etymology suggests that the 'thing' (*Ding*) is not removable); at the same time, the presence of ruins always points to other times, and, accordingly, other sites. Consequently, to return to Edensor's observations, spaces of danger suggest safe spaces elsewhere, spaces of delinquency suggest spaces for sanctioned activities, spaces of ugliness suggest spaces of beauty, and spaces of disorder suggest spaces of order. All of these categories entail a positioning of some privilege – that is, that there *might* be safe spaces and that there *might* be activities that are not subject to the label of 'delinquency' for particular bodies. Indeed, the (ostensible) disorder of the materials in ruins emerges as infringements on the orderly and bordered or bounded, especially regarding the body: risks of tripping, of cutting skin and of bruising muscles in unanticipated encounters with unexpected objects abound. Strange, unsavoury materials harbour risks of

---

⁷ Þóra Pétursdóttir's attention to the tensions enacted through heritage and ruins also draws on an understanding of ruin not as state or stasis, but as process. Heritage works to 'complete', she argues, but only insofar as the materials it curates to completion (or wholeness) are understood as somehow lacking. Of the 'concrete carcasses' of 'herring factories', Pétursdóttir suggests: 'Through their very presence, and the 'presence-effects' [reference is to Hans Ulrich Gumbrecht] they provoke, they can be said to utter their own critique of conventional heritage conceptions and urge us to critically consider the significance of the *concrete* and tangible in relation to heritage value, and of ruination as not only a negative but also a generative process' (Pétursdóttir 2013: 33–4, emphasis in original).

infection, of reaching under the skin or through the bronchi and bronchioles, enveloping the outside into the body. Spores, unseen but maybe inhaled, find new homes – the biting acidity of decomposing biota enters the body as particulates through the nose. The body is exposed to the elements, rendered vulnerable to abject objects.

The ruin is thus rendered excess and waste. These qualities have inherent spatial and temporal dimensions. In his article, 'Waste Matter', Tim Edensor observes: 'In ruins, processes of decay and the obscure agencies of intrusive humans and non-humans transform the familiar material world, changing the form and texture of objects, eroding their assigned functions and meanings, and blurring the boundaries between things' (Edensor 2005a: 318). Edensor evokes the 'meld' to express the ways in which 'stuff' mixes with the array of objects, 'a meld of other stuff: clinker, plastic, piles of thick lime, cleaning agents, adhesives, grease, oil, pitch' (ibid.: 319). Such objects 'may merge with other objects' and form 'peculiar compounds of matter, aggregates of dust or rubble' (ibid.). Fungus, rhizomatic and enigmatic, settles, he notes, 'using up matter and infolding it into itself' (ibid.).[8] The evocations of material melding together recollect the weird amalgamation of the plastiglomerate: concrete forms plasticizing, plastic becoming concrete. Discrete entities which were once objects in various stages of production and assembly lose their 'meaning' without the context of production and assembly and become material (again).

In a state of ruination, care is no longer taken to order such materials. The factory as a space of overtly organized material relations is, for Edensor, exemplary for the 'social-material order' of the 'imperatives of production' (Edensor 2005b: 98). This 'disorder' gives rise to an unruliness: Future eventualities are 'ripe for reappropriation' (ibid.: 108) and are present in the objects of ruins which emerge when human agency is reduced, that is, when processes of human curation, rescue and removal are absent, when they give way to decay, rust and other processes of material agency (cf. Edensor 2005a: 319). The materiality of the ruins thus gives rise to a different form of engagement that gives room for, and rise to, such processes of non-human agency.

Ruination exposes the formerly hidden: paint falls off, walls collapse or crumble, channels of energy/electricity/water/sewage are revealed and 'a compilation of varied materials through which the building has been organised bursts out of assigned positions in an efflorescence of deconstruction' (Edensor 2005b: 110),

---

[8] I have written elsewhere in more detail about the materialities of fungi (see Crane 2021a).

as Edensor puts it. Whilst some elements of ruin are reappropriated as objects, some are reappropriated as materials, that is, copper wiring is not retrieved from the ruin (or rubbish dump) for its 'thingness' as wire, but instead for the (economic) value that the copper can command *as* material. Whilst such acts of reappropriation do not leave the realm of economic value – and thus remains ultimately anthropocentric – such acts do stretch, even interrupt or break, the capacity of objects to remain objects. That is, as things decay, they lose their discrete forms, 'yielding to the processes which reveal them as aggregations of matter, erasing their objective boundaries', revealing 'those edges which could be felt and looked at' working against the ways in which 'the object was inviolable as a discrete entity' (ibid.: 114–15).

Ruins emerge as contradictory sites. They 'push back' on authoritative attributions of meanings: As Tom Nielsen argues, 'The concrete matter of the city will always exceed the ambition and attempts to control and shape it, and will always have features that cannot be exposed in the representations that planning has to work with' (Nielsen 2002: 54). It is, again, the close interpretation of the *materiality* of the ruin as site that reveals such meanings. For instance, there is, in Edensor's book and article, an evocative sense in which botanical and animal 'colonization' of industrial ruins suggests a shift in temporalities away from that of factory-based production, specifically with respect to how these 'intrusions' are evoked in terms of weeds and pests (cf. e.g. Edensor 2005b: 43–7). Specific species present in such locations show how crucial site is to determinations such as 'ornamental' or 'weed'/'pest'. In his article, he suggests,

> The spatial recontextualization and condition of objects in ruins draws attention to their material qualities, making evident the matter out of which they are made. This confrontation with the materiality of things can provoke a sudden awareness of the ways in which we are affectively and sensually alienated from the material world through the regulation of the sensory impact of things. (Edensor 2005a: 324)

The use of pesticides and planned eradication schemes in other sites (e.g. in agriculture) suggests that the presence of non-human species in such ruin sites emerges with and against human presences and aspirations. The emergent, unruly, presence of materials thus suggests the agentive properties of non-human livelihoods marginalized in anthropocentric worlds. The edges of the object at the site of the ruin reveal materiality, which is then catalysed by exposure into a trans-corporeal intra-action of *producing* the ruin.

Urbex: Practicing edges and photographic practices

Another mechanism through which the edges of the ruin are revealed is the practice of urban exploration. This is unscripted engagement with materials and sites and, crucially for Edensor's account of Britain (especially its larger metropolises), the absence of (electronic) surveillance (cf. Edensor 2005b: 21).[9] For Edensor, the practice of urbex (as I will abbreviate 'urban exploration' in what follows) traces 'the impact of the unruly affordances of ruins on bodies, and the subsequent coercion of people into entanglements and performances which disrupt normative understandings about what to do' (ibid.: 51). Here, I consider some urbex practices and artefacts for the way they produce specific discourses of ruin.

Whilst the photographic practices of urbex will be discussed at length later, this is not the only way such practices are documented and distributed. Paul Farley's and Michael Symmons Roberts's *Edgelands* is one example, a book. The book comprises of journeys to and amongst 'less-trammelled' landscapes, forgotten urban spaces: the space of the train tracks between Liverpool and Manchester, 'some of the most mature edgelands on the planet' (Farley and Symmons Roberts 2012: 236) – spaces left underdeveloped by development, the edges of modernization. The temporal duality becomes evident in the *materiality* of the site, problematizing the ostensible easy relations comprised in heritage sites (which, again, return later in the discussion of this chapter, cf. also Pétursdóttir 2013). Farley and Symmons Roberts argue:

> The heritage industry tends to rely on a kind of freeze-framing of time in order to present the tourist and visitor with a reordered, partial, tidied-up account of what happened at any particular site. Edgelands ruins contain a collage of time, built up in layers of mould and pigeon shit, in the way a groundsel rises through a crack in a concrete floor open to the elements. They turn space inside out …. Encountering the decay and abandonment of these places is to be made more aware than ever that we are only passing through. Edgelands ruins are unpredictable terrain.
>
> (Farley and Symmons Roberts 2012: 157)

---

[9] When Edensor celebrates the spaces of the ruin as giving rise to play, he explicitly acknowledges the advantage his gender and age give him (see also above). He is of a gender that these places might be considered safe (or, rather, not explicitly dangerous: the unruliness of the spaces does harbour some dangers), and his (middle)age works to exempt his activities, for the most part, from delinquency. It seems as if Edensor has overlooked the ways in which race shapes his being in and with certain environments: Activities rendered as play by people of certain ages and races will be interpreted as delinquency when enacted by different bodies, and the way these categories intersect gives rise to further interpretations (cf. e.g. Epstein, Blake and González 2017).

Edensor makes a similar, more succinct, point when he observes: 'Just as visiting ruins is a kind of anti-tourism, the ruin itself stands as a sort of anti-heritage' (Edensor 2005b: 139). This observation does not, however, stand in the way of Edensor taking photographs of his endeavours (like any (anti-)tourist).[10]

Photographing ruins (and the subsequent distribution of the images) has become widespread as a practice associated with urban exploration, a term comprising various practices of engaging with ruins, abandoned landscapes and, increasingly, other 'off-limit' urban structures such as high-rise construction sites.[11] These practices manifest in several forms for dissemination amongst practitioners and other interested parties. Most predominantly this happens through not only photography – online and in high-quality 'coffee-table books' – but also blogs, internet platforms and, more recently, academic monographs and articles.

The definition provided on 'Spurensammler' ('trace-collector' at lipinski.de) has the primary function not only of introducing the unfamiliar viewer to the collections of photos provided on the website, but also draws out several key issues pertinent to my discussion:

> Urban Exploration. An old hobby, with a (relatively) new name. Probably dating back to the Romantic period of the late 19th century, the last three-to-four years have seen increasing interest in the relics of the past, especially buildings, as part of our cultural identity. Even though contemporary explorers employ modern technology, it remains primarily a hobby of retrospection. A vision of the past.
>
> But if you look closely, you will notice some rules. Rules of decay that, in the future, will come to reign over everything that surrounds us at present. And, before long, the past becomes part of the future. We are, in effect, time travellers, and some of the stops on our travels can be seen here. (Lipinski n.d., my translation)[12]

---

[10] Edensor's images are somewhat different to many of the other urbex images in circulation: Edensor's are black-and-white rather than vibrantly coloured, and remain unlabelled. I find the lack of labels to the photographs in Edensor's work problematic for two reasons: Firstly, my unfamiliarity with the specific sites precludes me from exactly the kind of engagement he suggests the lack of labels will generate; secondly, I find the assumptions implicit in photographing sites of the UK and suggesting they are representative beyond their specific geographies to be universalist in effect (if not intent).

[11] The term has also been used by David Pinder to refer to activities such as drifting or avant-garde urban practices put forward by the Situationist International and was given a brief turn in the spotlight with the rise of 'psychogeography'. Accordingly, his use of the term 'urban exploration' appears interchangeable with 'psychogeographical explorations' (cf. i.e. Pinder 2005: 389). In this chapter, I am particularly concerned with those types of urban explorations that lead to ruins or underground tunnels and other 'off-the-map' sites (cf. e.g. also. Deyo and Leibowitz 2003 and Ninjalicious 2005).

[12] Original:

> Urban Exploration. Das Hobby ist alt, sein Name (relativ) neu. Wahrscheinlich schon verwurzelt in der Romantik des ausgehenden 19. Jahrhunderts ist in den vergangenen 3–4

Several themes emerge from this definition: Crucially, for this chapter, it suggests a link between material culture ('relics') and cultural identity – identifying in particular the explorations undertaken in the Romantic period as a precursor. By drawing on discourses of discovery, this part of the definition works to establish urbex as a legitimate discipline with a long heritage. Another key idea is the envelopings of different time frames present: 'old hobby' with 'new name'; 'relics of the past'; 'modern technology with 'a view to the past'; 'time travellers' and, most pertinently, the phrase 'the past becomes part of the future'. This idea ties in directly with the concept of *future artefacts* (elicited in detail in the introduction) – presences that announce (or betray, or confound, or depict, etc.) futures and other temporalities through their materiality. Finally, the idea of decay, as a process to which 'everything that surrounds us' is subject, consolidates the argument made in this chapter that considers ruin as both noun and verb. Such a definition of urbex, then, makes the practices very interesting for consideration in the rubrics of this chapter.

In Bradley Garrett's monograph *Explore Everything: Place-Hacking the City*, which traces his engagements with urban exploration as well as offering sociologically informed accounts of larger trends and developments, Garrett suggests:

> The most well-trodden avenue into urban exploration is through a fascination with ruins – buildings and places that have been left and are considered useless. Explorers seek out ignored and abandoned sites and photograph them as a sort of counter-spectacle to the contemporary city, where many people consider notions of 'development', construction and gentrification to be the normal course of things. (Garrett 2014: 8)

In this definition, even more so than the excerpt from '*Spurensammler*', the practice of photography is foregrounded. The visual spectacle is crucial to this definition, as well as the (quasi-)political acts of subversion entailed by appreciating the leftovers of the mainstream. Through such re-ascription,

---

Jahren wieder ein verstärktes Interesse an den Zeugnissen einer Vergangenheit zu verzeichnen, die mit ihren baulichen Ausprägungen Teil unserer kulturellen Identität ist. Auch wenn sich die zeitgenössischen Explorer durchaus moderner Technik bedienen, so ist es doch ein zuerst einmal rückwärts gewandtes Hobby. Der Blick in die Vergangenheit.

Aber wer hinsieht, stellt Regeln fest. Regeln des Zerfalls, der auch einmal alles andere erfassen wird, das uns augenblicklich noch umgibt. Und damit wird aus der Vergangenheit auch gleich wieder ein Stückchen Zukunft. Wir sind also quasi Zeitreisende und einige Stationen dieser Reisen sind hier zu sehen. (Lipinski n.d.)

furthermore, the processual character of ruination becomes evident and alternative practices of attributing meaning foreground the arbitrariness of sites being identified *as* ruin.[13]

For Garrett, the practices of urbex do not end with highly aestheticized photographs. He posits photography of ruins as either a temporary stage through which people pass and leave behind on their way to a more sophisticated urbex practice or as a separate, even less authentic (cf. e.g. Garrett 2014: 79), form of urbex. Garrett's narrative follows a split away from the 'well-trodden avenue into urban exploration … through a fascination with ruins' where explorers 'seek out ignored and abandoned sites and photograph them as a sort of counter-spectacle to the contemporary city' (ibid.: 8) towards a countercultural practice engaged by those with the inclination and who, crucially, can afford it.[14]

Garrett consequently draws on the concept of 'edgework', as alluded to above, to characterize urbex. Following Stephen Lyng, edgework is a concept that 'allows us to view high-risk behavior as involving, most fundamentally, the problem of negotiating the boundary between chaos and order' (Lyng 1990: 855). Common themes of edgework include self-realization/self-actualization/self-determination (cf. ibid.: 860), the threat of injury or death, whereby the risky behaviour works to warp temporalities, specifically, condensing time. 'Focused perception', Lyng notes, 'also correlates with a sense of cognitive control over the essential "objects" in the environment or a feeling of identity with these objects' (ibid.: 861). The agency of the environment, which for Lyng is activated by the process of identification, is a crucial insight for a reading of the practice within the rubric of material cultural studies.

---

[13] Accordingly, Garrett also suggests that the practice 'rarely leaves the city vulnerable. What the practice *does* challenge is the underlying message of constant and immanent threat promised by neo-liberalism that is used to codify the urban environment for "safety" and restrict the range of acceptable activities' (Garrett 2014: 18, emphasis in original).

[14] Garrett notes that urbexers are '*largely* a group of white, middle-class men' (Garrett 2014: 20, emphasis in original) but not exclusively so and that 'obviously, in order to have the opportunity for these sorts of engagements with the city, one must be secure enough financially and have enough free time that investing the hours necessary to research and explore sites can be accomplished. More importantly, one also has to view these spaces as primarily areas for play and not, for instance, potential housing' (ibid.: 21). Missing from this characterization of the people inclined to participate in urbex are those who might perceive the 'unmapped', unsanctioned spaces of exploration as a threat, most obviously as spaces that are not 'policed' (either by the state or by mechanisms of social control). Not all people with the financial and time resources will be inclined to participate: He notes, for instance, that only approximately 10–15 per cent of London explorers are female; able-bodiedness as a criterion is perhaps so obvious that Garrett neglects to mention it; and other criteria carried with the body – such as age, gender identity (including non-binary or fluid identities) and (perceived) non-het sexual identity – might prove prohibitive to those who perceive a threat embodied *by* 'group[s] of white, middle-class men' (ibid.: 20).

Referencing Jane Bennett's work in *Vibrant Matter*, Garrett argues that urbex 'is about space as much as time, about the event of discovery as much as the accumulation of knowledge, about things as much as people' (Garrett 2014: 32). Garrett's descriptions evoke images not only of thickness, of layeredness, but also of the agency of places: 'When we allow a place to teach us about itself, we give it agency, we begin to build rich tapestries that enticingly rearrange images of the past' (ibid.: 44). In the 'tension' between 'innermost yearnings' and 'life beyond the constraints of the material world', writes Garrett, 'we find a darker component of an imagined ruined future, a Ballardian formulation of urban apocalypse where the remains of our everyday existence become the *archaeology of the future*' (ibid.: 53, emphasis added).[15] With its emphasis on (documenting) the material surrounds, urbex, in particular urbex photography, emerges as a practice of engagement with future artefacts. Following such definitions, urbex practitioners' fascination with ruins tends to work in a nostalgic mode. It looks back at the same time it looks at the present. Michel Foucault's genealogy is useful for thinking through this here, in the sense that it works through uncertain origins, imagined origins and a simultaneity of various histories (plural) (cf. e.g. Foucault 1972).

Urban exploration engages in this tradition, implicitly and explicitly. The archive of urbex attests to an engagement in several key academic disciplines and areas of knowledge, in particular history, preservation and heritage, cultural geography and, crucially, archaeology. Bradley Garrett accordingly suggests: 'In abandoned bunkers, hospitals and industrial sites, we found moments caught between the present and the past, confrontations that flared up with unexpected material traces. Often we felt like archaeologists, assaying surface material without deep excavation to analyse the character of places; researchers conducting a survey of affectation' (Garrett 2014: 32). The renderings of urbex as a specific creative practice thus emerge from an attentiveness to the contradictions, both conceptual and material, of urban spaces.

In some images, such enveloping of time emerges with an even greater level of complexity. As an example, take any number of photographs of the Sanzhi project from Taiwan. One such image is 'Abandoned Future IV' by Lin 'Cypherone' Yung-Jie, which is still viewable on flickr and Instagram: The

---

[15] 'Urban exploration offers an important model for the intersection of history, present-day experience and the future' (Garrett 2014: 57). The enveloping of time-scales recollects that noted in the '*Spurensammler*' definition cited earlier. The works of J. G. Ballard, alluded to with the adjective 'Ballardian', will be examined later.

image has traces of filters – the saturation appears to have been altered, and the contrasts softened towards the edges. The now-destroyed village – Sanzhi Pod City, Taiwan – is paradigmatic for the idea I wish to evoke here: a moibus-strip present image of a past imagining of a future that never happened (nor can happen: the village has been destroyed). A similar image graces the cover of Alessandro Biamonti's *Archiflop* (with the almost flippant subtitle *A Guide to the Most Spectacular Failures in the History of Modern and Contemporary Architecture*). The 'UFO houses' (Biamonti 2017: 44) in fact combine the use of both materials central to this book – plastic and concrete. Tong Lam's volume *Abandoned Futures* similarly plays to the multiplicity of the idea, through the insistence on the plurality of future*s*. As erstwhile colleague Florian Freitag once explained to me, retrofuture is the most stable version of the future that you can get (cf. also Carlà-Uhink et al. 2023: 312).

As Biamonti observes, the structures look 'like something out of a 1960s-1970s sci-fi film' (Biamonti 2017: 44). Concrete, as the material of an imagined future, mobilizes in such images a *past* vision of the future to elicit a nostalgia for *another* present (not *the* present), a utopia that has never come-into-being (see Figure 5.1). In *Concrete and Culture*, Adrian Forty notes 'a

Figure 5.1 '三芝飛碟屋 – panoramio (1)' by The Erica Chang (CC BY via wiki-media).

long-standing association between concrete and utopian movements of all kinds' (Forty 2012: 8), citing a brief passage from Thomas More's *Utopia*. Concrete is, he suggests further, '*modern*. This is not just to say that now it is here, when before it wasn't, but that it is one of the agents through which our experience of modernity is mediated. Concrete tells us what it means to be modern' (ibid.: 14, emphasis in original).

The re- (or over-)working of urbex photographs renders the material layers of the depicted objects more tangible. The increased use of contrast not only stresses the differing moods of the sky, but it also emphasizes the signs of decay: the slimy sludge on the curves of the pod, the lichen present on the concrete stilts and the abundance of grass (or weeds) at the foot of the structure. The intra-action between industrial material and organic processes of decay is highlighted through the image rendering (e.g. through photoshopping). There is a sense of abandoned future that pervades through the renderings of the photograph; it is suggestive, I think, of a future artefact.

The use of colour is crucial in such images. Contrast is overtly rendered in the 'hyperaesthetics' of much urbex photography, where the un-smooth – peeling paint, cracks in concrete, jutting bricks, jettisoned materials – is highlighted.[16] The saturated use of colour can be seen as an incursion into the contradistinction of the functional muted 'pastel and mineral shades' (Edensor 2005b: 72) of 'backgrounded urban spaces' by using the 'offset' of 'brighter incursions of colour' (ibid.) generally associated with advertisements and displays with their concomitant interpolations of consumerism. The *lack* of work or care evident in the spaces of urbex – the lack of cleaning, polishing, sweeping, wiping, infilling of cracks and so on – becomes overt rendering work in the photograph, where the work and care is part of the artistic processes of layerings, manipulations and pixel-level shifts. Photoshopping, in such urbex photography, is not the 'smoothing over' of blemishes prevalent in glossy magazines, advertisements or profile photographs in social media (to name some more obvious sites), but is employed to specifically 'un-smooth'.

Whilst other images of urban decay, exploration of relics and other objects of interest to urban explorers are also available, the overtly edited photographs are the ones that attract my attention here. Photoshopping is one form of a number of digital manipulations that result in HDR images (also known as HDRI). HDR

---

[16] The evocation of 'edgework' to describe the practice of urbex as discussed with Lyng via Garrett earlier can also be used as a heuristic device to trace the 'edginess' of the photographs that document or celebrate the practice. Lyng himself suggests that 'the sense of the edgework experience as a kind of "hyperreality"' (Lyng 1990: 861).

is an acronym for High Dynamic Resolution Imaging, basically photography that attempts to recreate the luminosity that human eyes can perceive by way of digital manipulation. Numerous how-to guides are available online, and a book called *Urban Exploration Photography: A Guide to Creating and Editing Images of Abandoned Places*, written by Todd Sipes, includes sections on being prepared, how, when and what to shoot, as well as an extensive section on editing (with specific software tools, and 'finding your style', as well as information on settings and so forth). That book, and the plethora of how-to websites that precede it, testify to a mainstreaming of urbex (at the same time affirming Lyng's assumption of increasing involvement in 'edgework', cf. Lyng 1990: 882–3) as well as to increasing popularity of digital photography. This increasing popularity is perhaps traceable to devices becoming more affordable, but is probably also linked to burgeoning interest in sharing visual media online (i.e. through flickr, Instagram and on social media websites).

In the categories established at the outset of the chapter, then, urbex photography reveals a predilection for ruins and ruining. It is this jarring or crumpling (folding sounds too neat) of the future into a present over-determined by the material remainders of the past – rendered aesthetic through the visual form of photography – that is suggestive of the memorial function of the ruin. Such ruin, as noun *and* verb, is indicative of the shifts in scale and perspective I elicit through the concept of the 'future artefact'. The hyperaesthetics of urbex photographs refigure junk, rubbish, empty buildings and ruins as sites for imagining and as sites in discrete stages of 'becoming'. The activities of urbex that give rise to the photographs (independent of attitude towards the 'care' taken by extensive photographic manipulation) refuse to be contained by the constraints of legality. The images also show objects refusing the restraints of their object-ness, either through manipulation of the colours or through the use of colour to foreground the uneven surfaces, their materialities and, furthermore, the blurring of epistemological categories.

Toxicity, containment and other future projects: Nuclear temporalities and concrete materialities

Concrete as a manifestation of power is a crucial component of Chapter 3. There, I address the entanglements of materials with the production of power, specifically hydroelectric power and the power of the (modern) nation state. Concrete has a crucial role in the generation of another form of power, also specifically modern and particularly contemporary: nuclear power. Urbex encounters with concrete structures of abandoned nuclear power stations and the structures erected to hold spent nuclear fuel rods address the manifestations of ruin, contain(ment) and support through the material forms of nuclear power or, more specifically, nuclear infrastructures. Cymene Howe et al. suggest that whilst infrastructure has a 'future orientation' and 'generative impulses' (Howe et al. 2016: 560), it must be recognized that 'in many, if not most, cases we live and work among various kinds of ruined or faltering infrastructure' (ibid., cf. also Larkin 2013).

The archive of urbex images, then, might provide an avenue for considering and critiquing the processes and buildings that comprise ruins, and, crucially, our own implication in them. That is, also the ways in which depicted objects might object, might jar with expectations and might render the 'undesirable' highly aesthetic. Concrete materials – where concrete is both the opposite of 'abstract' *and* the building material, with its etymology suggesting to 'grow together' – also afford ways of 'integrat[ing] these two perspectives', that is, following Hodder 'to explore how the objectness of things contributes to the ways things assemble us, and to examine how our dependence on things includes the desire to be shorn of them' (Hodder 2012: 14). Urbex photography can be seen as itself enacting an exploration of the processes of thingness and objectness, through its archival function,[17] opening up engagements with (neglected) landscapes, buildings and rubble. The visual foregrounding of processes of decay can hence elicit a comprehension of ruin as both noun *and* verb.

Concrete is both a crucial component of the construction of nuclear power plants and also in the containment and storage of wasted nuclear fuel. This includes intentional storage, one aspect of the analysis that follows, as well as 'unintentional' storage, or the mitigation of contamination following nuclear 'accidents', such as the events of Chernobyl from 1986, or more recently, Fukushima in 2011.[18] When used as a material for the containment of threat, concrete is employed to mitigate *against* ruin. The use of concrete in a variety of structures to 'weather' against different kinds of threats is overly confident – utopian – in its

---

[17] For Michel Foucault, the concept of 'archive' does not just reference the material, the collection of texts and other artefacts, but also refers to the way in which this kind of information is organized, and how it emerges as a kind of 'discourse'. The way we talk about the archive is as integral to what constitutes the archive as the actual matter in it. This means that the kinds of questions we ask of it, and our specific imaginings of it, become just as crucial to the archive as the documents (or other artefacts) it stores (cf. e.g. Foucault 1972: 129).

[18] On the false premise of nuclear 'accidents', see Joseph Masco's deliberations on fallout, for example, Masco (2021).

belief in the material stability of concrete: the concreteness of concrete emerges as an aspiration, a goal, more than a reality. Concrete containment is, as I will show, conceptually and materially porous. This porosity pervades the distinction between 'conceptual' and 'material', integral to the double meanings of concrete.

The so-called 'Sarcophagus' at Chernobyl has become a site (if not a 'Mecca') of 'anti-tourism'. Andrew Blackwell's *Visit Sunny Chernobyl*, which was also mentioned in Chapter 2, gives it prominence, also making it the first site he visits in the book. Indeed, urbex photographs of the site are abundant. On visiting the Chernobyl Museum in Kiev, Blackwell describes an exhibit in one of the two halls ('a bizarre temple-like space', Blackwell 2013: 4) thus:

> Soothing Russian choral music emanated from the walls. In the center of the room lay a full-size replica of the top face of the infamous reactor. A dugout canoe was suspended above it, heaped with a bewildering mixture of religious images and children's stuffed toys. I tried to understand the room's message, and could not. Empty contamination suits lingered in the shadows, arranged in postures of bafflement and ennui. (ibid.)

Blackwell's declared lack of comprehension at the exhibit is linked to an excess of potential meanings that remain undecipherable. Without explanation or engagement, the exhibit's objects (are) simply object(s).

Elsewhere, Blackwell's description of the site pays some attention to concrete: the paths, the buildings (including the 'pectopah' (ibid.: 24), a recreational centre), some poles. Most obviously, concrete is the material of the 'Shelter Object', known as the Sarcophagus, which was built over reactor #4 after the events of April 1986. An 'Exclusion Zone bureaucrat' (ibid.: 29) named Julia explains that another newer structure, 'New Safe Confinement', is scheduled to be built over the original Sarcophagus:

> The New Safe Confinement, if it's actually built, is intended to last 150 years. The reactor building, though, will be dangerous for millennia. So maybe there will one day be a shelter for the shelter for the Shelter Object, and then a shelter for that, and we will continue down the generations, building – shell by shell – a nest of giant, radioactive Russian dolls. (ibid.: 31)[19]

The endearing image of the Russian dolls aside – an image that jars with Blackwell's own account of the desolation rendered through the abandoned

---

[19] References to the viewing of a World Cup soccer match suggest Blackwell's visit was in the northern hemisphere summer of 2006. The EBRD reports that construction was completed in 2018 (cf. EBRD n.d.).

objects of Kindergarten No. 7 (evoked through the contrast with a series of photographs that show children at play in the kindergarten itself) – a futurity of concrete containment is evoked. The concrete structure is evidence of, and in no way immune to, ruin: A malevolent ruin emerging through toxicity (cf. also Chapters 2 and 4), perpetually in need of further containment.

The term given to the concrete structure, sarcophagus, alludes to the stone coffins used, for example, by people of the 'ancient civilizations' of Egypt, Rome and Greece. It designates a burial place and a place of memory. The trope of monuments marking death used to designate concrete structures for the containment of toxicity is one that also emerges at another site, the concrete dome atop a nuclear waste site on Runit Island (in the Enewetak Atoll, part of the Marshall Islands in the Pacific). There, the image of the concrete dome as tomb is pervasive. Kathy Jetñil-Kijiner's poem and accompanying video 'Anointed' asks, 'Will I find an island or a tomb?' as she navigates the seas of the Pacific towards the concrete island, a 'concrete shell that houses death' where concrete emerges as 'solidified history, immoveable, unforgettable' an image rendered monolithic, particularly through the contrast with evocations of another shell, the turtle shell of Letao, a mythical creature (Jetñil-Kijiner 2018).[20] Across Jetñil-Kitijiner's work, nuclear threat and containment are brought explicitly together with considerations of climate change induced rising sea levels, and, by extension, the Anthropocene.

Concrete's *incapacity* to shelter, protect, even cordon off the nuclear threat, strips it of its perceived impermeability: The functionality of the material is shown to be interdependent on context, site, environment, from Chernobyl to Enewetak Atoll, Europe to the Pacific. Concrete is both 'thing' and 'object' and also the interplay of thing and object.

Despite these shortcomings – the dome is unable to contain radioactivity, the sarcophagus is in perpetual need of replacement – concrete remains one of the key materials used to harbour nuclear waste. The temporal dimension is compounded by spatial considerations: The material can only proffer stability over a (long) period of time if and only if the site is also stable. Vincent F. Ialenti evokes 'deep time' (following author Gregory Benford and science historian Martin Rudwick) and the 'long now' (via Stewart Brand) to address concerns of the long duration of the toxic future artefact. He suggests that the site of

---

[20] Christina Aningi, a teacher at Enewetak's school, has been quoted stating, 'We call it the tomb' (Willacy 2017). Mark Willacy employs tropes of vulnerability and exposure to describe both the concrete-topped dome on Runit Island and the lives and livelihoods of the human population on nearby Enewetak (for more on the 'Nuclear Pacific', cf. DeLoughrey 2012).

the controversial Yucca Mountain Project in Nevada, and by extension, other high-level radioactive waste repository sites, are sites 'in which distant future societies, bodies, and environments are engaged – in which relations between the living societies of the present and the unborn societies imagined to inhabit distant future worlds are made and remade' (Ialenti 2014: 29–30). The physical realities of the site are then only one component of the future artefact that is a nuclear waste repository: Crucially, the imaginative component must also be addressed and negotiated with and through the site.

Nuclear waste storage, accordingly, is a particularly pertinent example for contemporary negotiations of future artefacts: The waste must remain contained, even hidden, in order for its toxic effects to be mitigated. For a species as inherently curious as human beings, this poses quite a challenge. The point is, here, that the materials contained by the dome atop Runit Island in Marshall Islands, the sarcophagus at Chernobyl, or the concrete and salt structures that house radiated materials at the Yucca Mountain and at the Waste Isolation Pilot Plant in New Mexico (USA), or the underground cave at Onkalo in Finland, will remain radioactive for a long time. Concrete ostensibly acts as a container, a restrainer, arresting the toxicity of the radioactive materials. However, as with the case of the sarcophagi of Ancient Egypt, for instance, which despite warnings were opened by curious visitors (archaeologists in the interest of 'furthering' knowledge?), there is an urgency of ethics towards future generations that draws attention to the limits of collective memory, even linguistic capacity, to address the risk of exposure.[21]

Linguists, science fiction writers and other individuals with an array of backgrounds assembled in the framework of the Futures Panel in order to determine ways of best communicating the toxicity and danger inherent in the disposal of nuclear waste in the United States to future generations. The designers and engineers responsible for the Onkalo project in Finland have, in contrast, decided to not mark the site. As Andrew Curry reports, the reasoning was thus: 'If we mark it, we'd most likely invite people to look and see what's down there', in a quote accredited to geologist Timo Äikäs. 'Instead', Curry explains 'they've designed Onkalo to be as inconspicuous as possible' (Curry 2017: 57).

The Onkalo site thus encapsulates the issues of containment and communication, suggesting the 'edges' of that which can be communicated to the

---

[21] Cf. also Robert Macfarlane's chapter 'The Hiding Place' in *Underland* or David Farrier's chapter 'The Moment Under the Moment' in *Footprints* for explorations of the sites entailed in the containment of nuclear waste as well as the (im)possibilities of trans-millennial communication such practices entail.

unknown – future – addressee. As the Runit Island example also suggests, concrete can only contain to the extent that it cannot be broached – and rising sea levels will compound the inequalities of externalization present in the Marshall Islands. There is a disjunct between the spectacular potential 'fallout' (cf. Masco 2021) of nuclear fuel, on the one hand, and the often-unarticulated presumptions of a stable long duration of the materiality of the concrete structures that are used to contain them, on the other, around which such concerns coalesce. Insofar as concrete containment elicits a sense of deep time, to return to Ialenti, this deep time is not a depth of time that can remain buried. Here our conceptual framework of considering the past as deep, as something that can be buried, pushes back against the depths, and structures, of Onkalo, Runit, Chernobyl and Yucca Mountain. Concrete structures for containment of nuclear waste are 'future artefacts', and, despite the archaeological allusion to 'things that are buried', their capacity to interrupt the present and the future are integral to their material formations.

High-rise: Concrete hopes, concrete ruins

J. G. Ballard's 1975 novel *High-Rise* describes a different kind of container (as the novel's name already suggests). The structure of the eponymous high-rise is a rather obvious metaphor for stratified societies, with the lower (middle) classes occupying the lower floors and the upper classes the upper floors. The inhabitants of Ballard's imagined tower know this: there is talk amongst the upper floors/classes, at one stage, of mobilizing this stratification by 'balkanizing' the middle floors in order to secure amenities and security (cf. Ballard 2014: 126). The spatial metaphors of class are omnipresent: the high-speed elevators servicing the upper floors allow these inhabitants to bypass the lower and intermediate levels of the tower; access to the spacious roof is restricted, with its seagulls (birds associated with the leisure spaces of the beach) quite literally shitting on the lower floors; power cuts inevitably hit the lower floors/classes with greater intensity and regularity than the higher floors; the prime parking spaces closest to the building are allocated to the inhabitants of the upper floors, who can thus traverse common spaces much more quickly. And, finally, readers versed in UK events might draw connections between the social collapse of Ballard's fictional world and the real-world (partial) collapse of Ronan Point in 1968.[22] Today's

---

[22] Sebastian Groes suggests, '*High-Rise* ... voices concerns about the utopian but misconceived and under-funded tower block projects in the 1950s and, in particular, 1960s, which had come to an abrupt halt with the partial collapse of Ronan Point on 16 May, 1968' (Groes 2012: 123).

reader might think of the disastrous fire at Grenfell in 2017.[23] Both occasions are events, in the sense that they are material eruptions of otherwise slow processes of (social) decay and compartmentalization, that is, ruin, for which the social disintegration depicted in the novel provides an apt metaphor.

Concrete's influence is present throughout the building: It permeates all levels. Certainly, the weight of the concrete building pushes down on the lower floors with more intensity, and so the social decay portrayed in the novel – the ruining of social livelihoods, the violence (particularly towards women and children, and dogs) – emerges as a function of the building itself. However, an interpretation that stops at that metaphorical level will forgo some of the implications of the materiality of the world depicted in *High-Rise*.

A lot of the tension in the novel arises from concerns with the materiality of the high-rise. Laing observes, rather early in the novel:

> However reluctantly, he now had to accept something he had been trying to repress – that the previous six months had been a period of continuous bickering among his neighbours, of trivial disputes over the faulty elevators and air-conditioning, inexplicable electrical failures, noise, competition for parking space and, in short, that host of minor defects which the architects were supposed specifically to have designed out of these over-priced apartments. The underlying tensions among the residents were remarkably strong. (Ballard 2014: 17)

Material concerns, here and elsewhere, press onto and into the lives of the residents. This tension is a key component of the relations of the characters of the novel amongst themselves and instrumental to the development of the plot. These tensions, however, are depicted as being intrinsic not only to the relations between humans, but also between humans and matter, and between matter and matter.

One particularly conspicuous manner in which this is presented in the novel is through the various uses of the word 'break'. The word 'break' is configured as breaks in materials (i.e. glass), in the 'social fabric',[24] characters 'break into' things

---

[23] The film adaptation of *High-Rise*, directed by Ben Wheatley and released in 2016 after showings at several film festivals in 2015, ends with a voice-over of Margaret Thatcher delivering a statement on the state and capitalism – 'Where there is state capitalism, there will never be political freedom' – 'paving' the way for the social changes and neoliberalist practices that emerged through the 1980s. The fictional world melds with the real world here through politics and their material manifestations in the high rise.

[24] The term 'social fabric' itself is material: The word 'material' also has the definition of 'cloth, fabric', reinforcing this link. And it brings a lateral connection to the workers and labour practices at Rana Plaza and other factories with it.

(i.e. flats), there is the 'breaking up' of materials and, hypthenated ('break-up'), relationships. Undesirable elements 'break through', as in 'deep-rooted antagonisms that were breaking through the surface of life within the high-rise at more and more points' (ibid.: 33). Further examples include 'outbreak' (i.e. of hostilities or violence) and 'breaking out' (of spaces, with its conceptual link to spaces of incarceration), as well as 'breakdown' (with respect to social relationships, services like the elevator and, for instance, the breaking down of doors). Breaks in material, and breakdowns in services, are simultaneously configured through breaks and breakdowns in the material reality of the high-rise. The breaks, left unmitigated or in a state of disrepair, lead to a ruin that seems (almost?) willed.

'The steady accumulation of rubbish is symptomatic of an eroded boundary between "inside" and "outside"' (Viney 2007), asserts William Viney, drawing strongly on the theories of Mary Douglas and Michael Thompson to examine waste in *High-Rise*. 'In *High-Rise*', he argues, 'Ballard questions the bold ambition of high modernist architecture. These building projects implicitly communicate concepts of cleanliness and waste management, furthering a techno-modernist form of social engineering. ... Progressive ultramodern housing rose from the derelict slums and industrial wastelands' (ibid.). In Viney's framework of a tension between waste and value (a binary structure inherent in the work of Douglas and Thompson), social behaviours are made manifest through the material practices of disposal: wasteland begets waste-land.

The novel parades a series of flamboyant waste-making gestures. Rather than being hidden away in the garbage chutes and plastic bags of 'normal' high-rise waste disposal practices, waste-making is flaunted as objects are thrown from balconies and onto the car park (where the more expensive cars are the easier targets, as the residents of the higher floors and more expensive flats are granted parking spaces closer to the building). The residents resist the removal of rubbish from the site:

> Presumably they held this rubbish to themselves less from fear of attracting the attentions of the outside world than from the need to cling to their own, surround themselves with the mucilage of unfinished meals, bloody bandage scraps, broken bottles that once held the wine that made them drunk, all faintly visible through the semi-opaque plastic. (Ballard 2014: 195)

The porosity of the ostensible boundaries between self and other is manifested through this neglect. The inhabitants' refusal to distance themselves from refuse emerges as symptomatic for particular relations to materials of deterioration as the relations *between* the residents themselves also deteriorate.

In this vein, Sebastian Groes considers J. G. Ballard a 'literary anthropologist whose work speculates on current social and cultural trends by imaginatively projecting them into extreme situations' (Groes 2012: 123), with a concern for 'the ways in which contemporary social relationships are mediated and distorted by new forms of urban space' (ibid.: 124). Here, I suggest that it is not only the 'deep' phenomena of these relationships, which Groes places in opposition to 'surface appearance' (ibid.: 123), but also the *materials* of the urban space. If the novels 'reveal ... late capitalism's brutal reshaping of the social and cognitive processes that determine everyday lives' (ibid.: 124), then this reshaping surely needs some 'stuff' through which it occurs.

The echoes of Le Corbusier[25] in the description of the high-rise at the outset of the novel (cf. Ballard 2014: 6) and also noted by Groes (amongst others) will be for those familiar with the architect enough to evoke the central materiality of this chapter. The architect's oeuvre is (almost) synonymous with the Modernist use of concrete, shifting away from the use of concrete as a hidden building component –structurally important, perhaps, but better left unseen – towards those practices that saw it as a material to be celebrated. Concrete, as substance or qualifier, is mentioned throughout Ballard's novel, in particular at the outset as the stage is set for the novel. The initial description of the high-rise is littered with the qualifier 'concrete': an empty ornamental lake is described as 'an empty concrete basin' (ibid.: 3), the area surveyed with its 'massive scale of ... glass and concrete architecture' (ibid.). Ballard has his protagonist Laing gaze across 'the parking-lots and concrete plazas below him' (ibid.: 5), and his initial sense of 'something alienating about the concrete landscape of the project – an architecture designed for war, on the unconscious level if no other' (ibid.) brings him to liken it to a 'row of concrete bunkers' (ibid.).[26] The high-rise itself is concrete, and the surroundings are a 'concrete landscape'.

Indeed, for our focalizer Laing, the 'concrete landscape' of the high-rise is alienating (ibid.: 5), evokes ambivalent feelings (ibid.: 28) or is concomitant with

---

[25] 'A house is a machine for living in' wrote Charles-Édouard Jeanneret, better known as Le Corbusier, in his *Towards an Architecture* (Le Corbusier 1986: 107). Often regarded as one of the key Modernist architects, Le Corbusier's philosophy has been influential beyond the discipline itself. This quotation is rendered somewhat differently in J. G. Ballard's *High-Rise*, but the central tenets remain: 'The high-rise was a huge machine designed to serve', the narrator of *High-Rise* notes (Ballard 2014: 6).

[26] John Beck uses the phrase 'womb- and tomb-like' (Beck 2011: 82) to stress the bunker's ambivalence. He notes further: 'The disturbing and relentless oscillation between life and death, ruin and rubble, nature and culture, exposure and concealment, image and object, art and atrocity, that vibrates deep in the rebars of the bunker's reinforced concrete is the tremor of modernity's ambivalence no amount of cultural recuperation can, or should, contain' (ibid.: 98). For more on bunkers, see also Silke Wenk (2001) and Luke Bennett (2013). There is a whole bunker(ed) 'rabbit-hole' hiding behind this footnote.

intrusions into privacy, evocative of an Orwellian world of surveillance: 'People who were content with their lives in the high-rise, who felt no particular objection to an impersonal steel and concrete landscape, no qualms about the invasion of their privacy by government agencies and data-processing organizations, and if anything welcomed these invisible intrusions, using them for their own purposes' (ibid.: 44). The associations evoked by concrete oscillate, like the project of the high-rise itself, between the utopian and dystopian, inevitably ambivalent and, increasingly, oppressive: 'He [Wilder] was constantly aware of the immense weight of concrete stacked above him, and the sense that his body was the focus of the lines of force running through the building, almost as if Anthony Royal [the fictional architect of the high-rise] had deliberately designed his body to be held within their grip' (ibid.: 62). The turn of phrase, using the keyword 'design', forges a further link to the Le Corbusier quote; the qualification 'deliberately' suggests both care and a slow, ongoing process. Throughout the novel, concrete has an agency in excess of that of a dormant construction material: Its capacity to ruin is portrayed as an active, malignant, even menacing (cf. ibid.: 144), affective force: 'Their real opponent was not the hierarchy of residents in the heights far above them, but the image of the building in their own minds, the multiplying layers of concrete that anchored them to the floor' (ibid.: 77).

The concrete of the building enables construction and gives 'rise' to the high-rise, and, at the same time, facilitates the fall into decay, both social and material. This is social and material relations in ruin:

> Even the run-down nature of the high-rise was a model of the world into which the future was carrying them, a landscape beyond technology where everything was either derelict or, more ambiguously, recombined in unexpected but more meaningful ways. Laing pondered this – sometimes he found it difficult not to believe that they were living in *a future that had already taken place*, and was now exhausted. (Ballard 2014: 208, emphasis added)

In *High-Rise*, the force of concrete exudes tensions resulting in breaks, both in the material and in the 'social fabric'. Materiality is constitutive to social relations – the social *structures* that uphold the high-rise –and lead to its ruin. The myriad of breaks suggests that the concrete pervading the fictional world, once formed by human interventions, now performs an agential pressure to form interventions into the human. Materiality blurs into mattereality.[27]

---

[27] This neologism is not my own, but rather I borrow it from a workshop held at the University of Freiburg, organized by Juliane Schwarz-Bierschenk, Miriam Nandi and Ingrid Gessner in March 2017.

Conclusion: DisplaCEMENT

Concrete entails displacement, at large scales. In Chapter 3, I looked to concrete structures – dams – which displaced water and with it peoples and livelihoods. Concrete, through its construction, also entails the displacement of sand, at increasing rates. Jan Zalasiewicz observes:

> Human-made rocks are everywhere. For sheer bulk, concrete now reigns supreme; we have manufactured something like a half a million metric tons to date. That is about a kilogram of concrete for every square meter of the earth's surface. Concrete forms the superstructures of our buildings, roads, and dams …. It is already a signature rock of the Anthropocene. (Zalasiewicz 2016: 34)

Concrete's impact on the environment extends beyond its infrastructural uses: Cement production accounts for some 5 per cent to 10 per cent of carbon dioxide emissions and each ton of cement finds a correlate in a ton of carbon dioxide (cf. Harkness, Simonetti and Winter 2018); concrete is, by way of cement, transforming – displacing – the oxygen we breathe.

Concrete also entails containment, in the quote above it figures obliquely as a kilogram that could cover each square meter of the earth's surface. The image does not seem strange writing here, in the city, where concrete accumulates, covering the ground and stretching into the soil and up into the air. But if I imaginatively travel to other sites I am familiar with – the farm and bush of my childhood, the deserts, the beaches, the forests, the mountains and other places I have come to know since – the image is 'out of place', displac*ement*.

In the structures that are supposed to mitigate radiation discussed in this chapter, or in the 'machine for living' evoked in J. G. Ballard's *High-Rise*, concrete contains. *Concrete does concrete things*. But its omnipresence as infrastructure does not mean it is dormant, external or stable. *Concrete is not always concrete*. The urbex images of concrete ruins and the structures of Laferrière's memoir evoke an aesthetics of ruin, an aesthetics of crumbling infrastructures that cannot contain (any longer).

The fascination of urbex with ruin is, I argued, also a fascination with timescales and agencies that defy the human. By evoking *ruination* alongside *ruin*, Laura Ann Stoler suggests that 'ruination is a corrosive process that weighs on the future and shapes the present' (Stoler 2008: 194). In this way, ruins extend along scales of time, extending from the (far) past and into the (far) future. And so, the figure of the future archaeologist returns. Waste dumps are their playground, or as David Farrier quotes Michael Gillings: 'Our landfill would be their goldmines' (Farrier 2020: 261, cf. also Rathje and Murphy 2001). In his future imaginary of our 'modern landfill sites', Farrier summons 'sealed-in leachates, rich in calcium carbonate from buried concrete', which 'may eventually produce a kind of cement to bind the clutter of household rejects, fly-ash particles and soiled packaging into a solid mass' (Farrier 2020: 261). The future of ruins, in this evocation, is the present of Kamilo Beach, Hawai'i: a plastiglomerate cousin, a member of the family of anthropogenic strata, melds of materials. We are back with *plasticoncrete*.

6

Conclusion

Thinking through materialities

Plastic is pervasive

It's in my life: my wardrobe, my kitchen cabinet and refrigerator, my toolbox, my bathroom, my computer, my transportation (bike, train) and on and on. It is increasingly in my body. It's probably in yours, too. Studies such as Kieran Cox et al.'s 'Human Consumption of Microplastics' suggest that we ingest thousands of microplastics daily.[1] Plastic is consistently mentioned in all kinds of media sources as an environmental issue, alongside crises like climate change. Plastic waste, references to what I called the 'Plastic Pacific', products made of recycled plastic, marketing ploys of 'plastic-free'… Talk of plastic has increased dramatically over the time I have engaged in working on this book. I downloaded an app 'ReplacePlastic' to my smartphone, for instance, that allows me to scan products that have 'superfluous' plastic packaging and the data is then (magically?) sent to the distributors, who are supposed to somehow respond to this information by using less plastic packaging; I can shop in a plastic-free supermarket, bringing my own containers ('Unpacking Plastic' by not packing with plastic); at some cafeterias, pre-Covid, it was possible to save 10c by bringing my own cup (and thus reducing plastic waste). Web pages promise ten ways I can reduce my (plastic) footprint or even embrace a plastic-free lifestyle; cities and entire countries that have introduced 'plastic bag bans'; the EU legislating to

---

[1] Cox et al. postulate that, although they state that their estimates are subject to variation and are probably understated, annual microplastic ingestion ranges from approximately 40,000 to 50,000 pieces, which approximately double when inhalation is accounted for; individuals who drink bottled water only can double this number again (Cox et al. 2019).

eliminate plastic straws; the list goes on. Our plastic practices manifest in rather *concrete* ways.

Concrete is more evasive

Than plastic, perhaps. And only at first. The qualifiers abound – as they are bound to – because if I think, deliberately, through my daily routines, the ways in which concrete shapes my life become increasingly noticeable.[2] It is not present so much as omnipresent. Concrete is in the paths, roads and highways, buildings, bridges, telecommunication masts and electricity line pylons, the railway sleepers, park benches, as well as in the underground tunnels transporting water and run-off below the surface, producing electricity with dams ('Megadam Materialities'). Concrete shapes – structures – my movements and privileges in very particular ways. One such way is as infrastructure, the basic physical (and organizational) structures and facilities that support my quotidian lifestyle, but remain *un*noticed. For Patricia Yaeger, 'infrastructure registers' in (literary) texts 'in its absence as dysfunction, nonpresence, decay', that is, as 'organizing structures [which] are present but barely visible' (Yaeger 2007: 16):[3] 'Concrete Ruins'. Concrete emerges as particularly *plastic* in its ways of shaping our lives.

Concrete coalesces in and as urban space, as infrastructure connecting spaces through (unseen, unfelt, unsmelt, etc.) castings. Plastic pervades pretty much everything and is persistent. For if we recognize the use of plastic in our day-to-day lives, perhaps urged by media accounts, and if we might be able to turn our attention to the concrete material practices (in the sense of 'things made of concrete'), we are still reckoning with the objects that comprise our interactions with our worlds. Thinking through materialities objects (verb) to objects (noun), suggesting that this truncates our engagement with our world(ing)s. It means moving beyond the lists, the concatenations of plastic and concrete: Thinking through materialities is thinking laterally, temporally and spatially, through relations and structures of relations.

---

[2] Noticeable, 'being known' (Chantrell 2002: 345). Or perhaps 'obvious' 'in the way' (ibid.: 350). Throughout this book, I have tried to prioritize verbs and adjectives pertaining to knowledge that do not reiterate a paradigm of sight as knowledge. Concrete and plastic are more than visual phenomena. So here, it's noticeable, not 'evident', for example, 'obvious to the eye or mind' (ibid.: 187) from *e-videre* ('out' and 'see'), nor apparent 'appear' (ibid: 26) or a host of other adjectives that draw on (or 'foreground') the visual paradigm in their metaphors of knowledge as light/sight.

[3] But it is the invisibility to which she returns, cautioning readers to 'remember to read this absence as a taking for granted of infrastructural privilege' (Yaeger 2007: 17). See also Larkin (2013).

Plastic as concrete; concrete as plastic. The heuristic of *plasticoncrete* uses the qualities of the terms, as materials, as adjectives, to shape inquiries into materialities through this book by means of 'cross-examinations'. The oscillation between the two materials in the longer chapters, from plastic to concrete to plastic to concrete, is a deliberate attempt to generate a sense for the ways in which these particular materials *in conjunction* give rise to a thinking through materials in the mode of the *plasticoncrete*. Thought as materials, rather than as a series of objects, concrete and plastic emerge separately and jointly to reckon with scales and infrastructures of relation.

The recent fashion for interior design objects made of concrete, like plant pots for instance, works to insert concrete as a material that *re*places plastic that in turn replaces other materials. In the early stages of working towards this book, I took the time to mix some concrete and mould it with plastic cups into a plant pot. The opportunity to engage in a concrete practise, to get my hands dirty, was also an exercise in waste production, as unfortunately nothing 'really' became of the pots. What it led to was a realization of the techniques of concrete production are not as straightforward as I assumed and it was also a reckoning with the industrial processes that are entailed by the production of plastic and concrete.

Plastic, too, is mobilized as a *re*placement of concrete or of the functions we might readily associate with concrete. Housing, for instance. The tarpaulin, or sheet(s) of plastic, are forms of shelter. From recreational activities for the privileged (like camping or thru-hiking, where it is admired for its weight) to the transitory shelters afforded to temporarily or permanently displaced peoples across the globe. My story of making only hints at the stories and stores of these materials: The entanglements of labour, logistics and the *longue durée* that lie[4] behind – and within – concrete and plastic. For whilst I might assume I can mix concrete, I cannot make cement, and whilst I could purchase a plastic 3D printer, I would still rely on the production of plastic pellets. Even at the individual scale, both materials remain industrial products.

To replace plastic bottles with bottles made of other materials – plastics with higher biodegradability, for instance, such as bioplastics – is to engage in the kind of plastic thinking that led to the (over)abundance of plastics in our lives. This mindset is what gave rise to many applications of plastic in the first place, a further

---

[4] Lie evokes here the double sense of Umberto Eco's question 'Where does the truth of ecology lie?' (Eco 1987: 49): a non-truth, here rather not-quite-entire truth, as well as an emplacement (or, to push it a bit further, empla*cement*).

iteration of replacement narratives pertaining to material practices. It is not the *materials* that require replacement in a set of behaviours keyed to reduce impact, but rather the *practices*, the behaviours themselves. *Plasticoncrete*, the conjoined heuristic that comes from thinking these materialities in conjunction, is about infrastructural flows as much as discrete objects. Systematic interventions, as much as individual inventions, are required.

Both concrete and plastic have liquid phases – recall that concrete is transported in its liquid state maintained by the perpetual motion of the rolling drum on the back of trucks – before they become manifest.[5] Stressing this quality of fluidity – which is key to functional infrastructure – is to stress the 'un-solidity', the *plasticity* of these otherwise *concrete* materials. This book returns again and again to the idea of materials stretching further than their manifestation as objects. Stretching is here both temporal (so reaching back into the past and forward into the future)[6] and also spatial, as the particular particles which comprise these materials suggest relations between seemingly disparate spaces. This is the sense that the cement provides uniformity between divergent sites as an industrial product, even if the aggregate is spatially specific (if not local, cf. e.g. Tweedie 2018), or that plastic derives from situated deposits of fossil fuels, is transported and fabricated until it becomes plastic, only then, after a (short) period of (human) use, to begin on a long journey of decomposition, which may or may not entail cycles of recycling. Future artefacts in the making.

All kinds of small-scale, quotidian practices have been comprehensively transformed by plastic (ultra-)awareness. Indeed, plastic is anything but small-scale. The list of interactions and sites of plastic above is just the 'tip of the iceberg', the gust in the cyclone, the plastic bag in the gyre. The 'Plastic Pacific' suggests as much. In this vein, Charles Moore and Cassandra Phillips argue:

> Plastic flotsam is the end product of an eons-long chain of transformations beginning with the planet's earliest lifeforms in the oceans … planktonic

---

[5] A lot of the theorizing done with globalization has used metaphors of flows. Whilst I most specifically discuss the materials of this book, concrete and plastic, in conjunction with discourses of modernities and Anthropocene(s), there is a crucial and pervading sense in which they are also materials of globalization. *Of globalization* both in the sense that their ubiquity derives from the processes and unevennesses of globalization, and also in the further sense that they (help to) constitute globalization. References to logistics (for instance, in my discussion of provenance stories via Marriott and Minio-Paluello, de Botton and Hamid in Chapter 4) and to political projects of modernization (as, for instance, in Chapter 3) corroborate this unarticulated subtext, as do the emphasis on the Pacific as an extra- or international site and on oceans as 'sinks' (Chapter 2), and the abandonment of factories in the Global North against the foil of collapsing factories in the Global South (in Chapter 5).
[6] Note the spatial component of this description of temporality (forward, backward), which maps differently across different languages.

creatures and algae living and dying over billions of years, blanketing the sea floor in numbers beyond our ken. The still-settling earth folds over on itself, trapping these vast deposits in pockets, pressure cooking them into viscous black gravy loaded with hydrocarbons. In a sense, our plasticized ocean represents recycling at its most epic, and worst. (Moore and Phillips 2012: 24–5, ellipsis in original)

The oxymoron of 'Synthetic Evolution', as he dubs this process, is revealed as less of an oxymoron and more of a tautology. The *longue durée* of plastic reaches back to 'first life' 'germinating billions of years ago': the 'black gravy loaded with hydrocarbons' reminiscent of Morton's erroneous but nonetheless useful 'dinosaur parts' (Morton 2010: 29). Plastic is oil is hydrocarbon is life. The epic scope of the history of petroleum-based synthetic plastics stands, ultimately, in stark contrast with the short-lived transience of the use of many kinds of plastic, even at a quotidian, human scale.

I suggest with my analysis of a cling-wrap advertisement that the materialities of plastic undergo a temporally shifting (plastic) slow-fast-slow process that map across the practices and becomings, both human and non-human. The slow-fast-slow maps, roughly, onto a pattern of composition-use-decomposition (see Chapter 4). The patternings of acceleration and deceleration of plastic become, when thought together with concrete, accruals of materials across the surface of the earth, coalescing around centres of modernization or 'development': urban, but also, for instance, industrialized agriculture. With *plasticoncrete*, these assumptions require some unpacking. The 'straightforwardness' of these assumptions regarding the plasticity of plastic and the concreteness of concrete is a function of etymological roots, but moreover a function of two different kinds of scale. These scales, again, are temporal and spatial.

Our entanglements with concrete and plastic occur on different timescales.[7] Our interactions with plastic are encounters with plastics in the kinds of objects concatenated in lists: This is the domain of the marvellous array of goods Yarsley and Couzen's 'Plastic Man' encounters in their 'Plastic Age' (imagined in 1941), the stupendous collection in the guts of Laysan albatross photographs, the 'Plastic' of Susan Freinkel's cover and her explorations of her *Toxic Love Story*

---

[7] Another footnote on my use of the plural first person 'our': Here, it oscillates between an 'extended I' – suggesting a specific positionality that reflects my own privilege but somehow exceeds it, however marginally – but also the way this imagined collective diffracts through a more general, and yet deeply problematic, 'universality', of humankind. We might all be in the Anthropocene, but we are not all in it in the same way, as Rob Nixon has succinctly put it (this is probably the last footnote on this topic).

(the subtitle to her book). These are everyday objects that we purchase and dispose of, that are of (much) shorter utility than the most tangible of timescales, a human lifetime. Analyses of objects can account for present or future wastes, as the object is, for instance, discarded and deteriorates, but have a harder time accounting for 'past wastes', or those excesses generated by extraction and production. George Marshall alludes to this problem in *Don't Even Think About It*: Climate change, with its policy emphasis on emissions, foregrounds present and future waste as a 'tailpipe' problem, Marshall argues, by backgrounding the waste of 'wellheads', or extraction technologies (cf. Marshall 2015). An under*lying* logic of externalization is at work here (cf. Armiero 2021, Patel and Moore 2020).

Concrete seems, in contrast, much less ephemeral. It, ideally, stretches beyond a human lifetime. If plastic emerges as the more pervasive, more tangible material, this is a consequence of this scale. Perceptions of our capacity to react to resource use seems to be correlated with on our understanding of our interaction with these resources: the concatenation trope, with its plastic predilection, suggests as much, and shifts of responsibility to the individual – like the idea of the 'ecological footprint', famously promoted by big oil – purposely reinforce this individualization.

Interactions with plastic take place, for the most part, on radically different timescales to our interactions with concrete. Our interactions – entanglements – with cling-wrap seem vastly different to those with, for instance, the university library building (or the Shard (Mark Miodownik's example) or the Burj Khalifa). Concrete is, after all, much more concrete, right? But only if we take ourselves as the reference point. In the timescales of the earth, the difference between several hours or days, on the one hand, and several decades (or even centuries), on the other, becomes increasingly negligible. In the timescales of the earth, human engagements, incursions and entanglements with and into the materials of the earth – 'the Anthropocene' – might scarcely warrant a distinction between the timescales entailed by plastic and concrete. The phase of utility of material practices – the time during which the harnessed resources fulfil some kind of (anthropocentric, useful-to-humans) function – becomes miniscule. This does not just apply to plastic, but also to concrete. The short duration of utility is a function of scale, which in turn is a function of relations. It is a compression, a coalescence and a coagulation – a messiness marked by an insistence on form.

Heather Davis's deliberations on the distinction between 'inheritance' and 'transmission' are useful for thinking through the unevenness of the spatial (and temporal) 'alterlives' (this term comes from Michelle Murphy) of materials (cf. Davis 2022, and also Chapter 2). Where inheritance suggests a relationship marked by passing on of benefits – often economic – transmission suggests the ways of being subjected to, making do with, forging livelihoods amongst the discards and ruins (see also Chapter 5).[8] Future artefacts, again, are not dormant, they exude toxins (see also Chapter 2) which effect differently situated communities in different ways: tenuously articulated plural positions (the 'we' and 'our' I try to footnote throughout the book) begin to crumble or break down; redress will require assemblages across identity and species borders employing strategic interventions and tactical ploys.

Geological change, Jussi Parikka argues in *A Geology of Media*, is punctuated by different temporalities; it is also open to folds. Parikka interrogates the notion of geology – in a sense, the study of earth's timescales – to suggest that stratification is 'not necessarily a smooth, slow process, but multitemporal, speeding up, slowing down' (Parikka 2015: 119). This is particularly useful as a means of imagining the material (after-)effects of materials. The image 'We Were Here Now' by artist Julia Sperling which graces the cover of this volume envisions such a stratification of materials, encompassing 'the mind-blowing scale' of plastic, concrete (and aluminium) as 'firmly rooted in the geological record' (Sperling 2018). Both Sperling and Parikka evoke the fossil, as well as the technofossil more specifically. The fossil, as a deposit of such processes and a surviving 'remnant' of material engagements, is a powerful trope to suggest the disparate timescales of humans and their manufacture. For Parikka, the technofossil and zombie media are evocations of paradoxes of materiality and technology, imbricated with paradoxes of archive and archival access (here access refers to technological means rather than (political) privileges). But for the arguments I make regarding the agency of materialities, this notion does not quite do the work I need. The 'fossil' is, after all, suggestive of a process of recovery, or discovery – its etymology comes from the Latin 'to dig'. It rests, dormant (or inert?), until it is 'rediscovered' or uncovered: dug out. In my articulation of the 'future artefact' and of 'invention', I preferred the 'making' of 'forms' as well as the idea of 'coming into relations' that suggest, etymologically at least, a more agential, active process of imagining materiality, one that does not necessarily re-centre the human, and, crucially, specific ideas of the human.

---

[8] Davis traces her biographical connection to, and economic benefit, from the plastic industry (as well as situating herself as benefitting from settler colonialism more broadly, cf. Davis 2022: vii–x). Stephanie LeMenager undertakes a similar autobiographical tracing with respect to her family's involvement in the oil industry, cf. LeMenager (2014: 1–2).

The future archaeologist – also evoked by Parikka,[9] and also found in other texts, for instance, in Heather Davis, David Farrier and Roger Luckhurst – is an interesting figure, one that emerges as integral, perhaps constitutive, of the Anthropocene. The future archaeologist also reiterates specific privileges: they have, we might assume, a certain kind of education, embody a certain status. But, as post-colonial and decolonial critics of the Anthropocene note, the 'future' itself maps only with great discord across spatial and temporal relations. In Kyle Powys Whyte's phrasing of 'Our ancestor's dystopia now', the foldings of temporal dimensions become more evident for the ways they break with linear or smooth conceptualizations of time and space. The present, Whyte cautions, is already someone else's – someone's particular – future. The repercussions of the Anthropocene – if it can be as singular as the definite article suggests – map unevenly across the globe.

Accordingly, our entanglements with concrete and plastic also occur at different spatial scales. Again, this is predominantly due to an anthropocentric tendency to see ourselves as the reference point. From the perspective of a human, the difference between the space taken up by the plastic bag, which we can hold in our hands, and fold into a package even smaller than our hands, and that of construction sites such as the Shard, the Burj Khalifa, or the stretches of road we might travel along or the (hidden) infrastructures of drains, overpasses and tunnels that shape our journeys through our quotidian lives, seems large. On the scale of the planet, such differences become increasingly negligible. This might account for our conceptual grasp on our capacity to act, react or interact with plastic, as opposed to concrete, and hence, perhaps, the deluge of plastic in the media representation of environmental issues, particularly seen in contrast with concrete.

And yet, the sheer size of the 'Plastic Pacific', disparate as it is, belies these assumptions. The scale of the individual, both the human and the 'object', is defied when thinking through materiality, reckoning with these kinds of structures that reach beyond discrete locations and times. In 'Scale Critique for the Anthropocene', Derek Woods argues that this is a function of the 'disjunctures and incommensurable differences among scales' (Woods 2014: 135). Woods, through his analysis of *Powers of Ten*, reminds us that 'zoom does not imply actual motion, but a changing resolution' (Woods 2014: 134) and that the

---

[9] Parikka picks up the plastic bag, by way of Chatonsky, to illustrate this trope: 'A plastic bag can last hundreds of years when I only have 2,500 weeks left to live. This disproportion between the human life expectancy and the one of our technical artifacts gives a new dimension to our time' (Gregory Chatonsky, as quoted in Parikka 2015: 120).

compatibility of scales suggested by the technique of zooming falls short of *representing* the real material differences that ensue from engagements with the world at different scales. Insects, Woods notes, can 'defy' gravity by walking on ceilings, a function of surface forces that are unfamiliar and untenable at the scale of humans (ibid: 136). Material worlds aren't (necessarily) extrapolatable from the scale(s) with which we are most familiar from our quotidian experiences.

For Mark Miodownik, scale-thinking is integral to material thinking: 'All materials have a common set of structures within them', he notes, thinking specifically about atoms (Miodownik 2014: 239). Extrapolating from this, he argues that similarities between materials become increasingly evident: 'In order to visualize this connection between all materials, we need a map of this Russian doll-like material architecture: not a normal map that shows the variety of terrain on a single scale, but a map that shows terrain on a variety of scales: the inner space of materials' (ibid.). I suggest that thinking through materialities entails one way of doing this work. Technical materialities – materialities of a however-defined modernity such as those of plastic and concrete – do not just extend the realm of the human (if it is possible to talk of such a singular figure), but push back *on* the human, extending the relations possible, and also truncating relations.

The idea of the compression, of the temporal and spatial limits of engaging with materials that *cannot* be excised from its larger networks of relation, is not just an exercise in intellectual play. The scales at which (radical) changes need to be enacted require thinking through – in the terms of this book, materialities – for the impacts of material practices already map across our own livelihoods. We need ways of imagining *how* this can be implemented, ways that go beyond individualizing frameworks of the 'ecological footprint' or indeed 'doing without' (acts of renunciation at the individual scale, like 'plastic-free years': there has to be fun ways of doing this, of being a 'bad environmentalist' (cf. Seymour 2018).[10]

---

[10] Abstaining, doing without, waiving, forgoing: The list of words constitutes *negative* relations, and – as Val Plumwood (1993) reminds us – thus structurally if not literally support the validity of that which is 'done without', 'forgone' or 'abstained from'. (It feels like a kind of bizarre diet: interested somewhat in the performativity of eating a shitty salad instead of something else with carbohydrates, fat or whatever the diet itself is 'doing without'. The German word *Verzicht* is lacking a good translation into English here.) In *Bad Environmentalism*, Nicole Seymour traces ironies and affects across media and cultural artefacts that engage with environmental thought in contradictory ways to help attune to the pervasiveness of this line of thinking. What we might need is more (individual, institutional) flexing of sustainable practices – I'm thinking of @iamharaldur's boasting on twitter of paying the second highest amount of taxes in Iceland. We might need to channel the energy of the subversive, like the individuals who would sip at their coffees for hours in German trains to circumvent Covid-19 mask mandates, or the child that licks a biscuit to claim it for themselves. We might need to be imaginative in reconstructing the frameworks of status in order to tackle

The impacts of materialities map unevenly across the globe, affecting others, both human and non-human, in vastly different ways. Rob Nixon's explorations in 'slow violence' are a case in point, as well as the variegated work that assembles under 'environmental justice', decolonial approaches to the Anthropocene or work at the nexus of critical race studies/critical animal studies and pollution. Thinking of the effects of the various ecological crises in terms of the future, as a temporal issue and through the scales of time, is an effect of countless privileges. It is as much a spatial issue, one that reverberates through scales of space. Imagining the souvenir – the Berlin Wall encased in plastic, the *plasticoncrete Denkstoff* I evoked in the introduction – in other times is also imagining the souvenir in other places: The temporal dimension is imbricated with the spatial. As the object stretches away from its present presence, it becomes increasingly *material*. Thinking through materiality suggests the way in which our material practices – as plastic (in the sense of 'adaptable') as they may be – have concrete repercussions.

Getting concrete, being plastic

My work on concrete and plastic, over many years, has shifted my comprehension of the world with which I interact on a daily basis much more than I anticipated. It has resulted in some purchases and some 'phasings out' that are predicated on my position of privilege, particularly as pertains to the continuing inclusion of plastic artefacts in my life; my grappling with plastic toxicity traced in Chapter 2 cannot be undone.[11] It has engendered a respect for the limits of my own knowledges – my limited knowledge of chemical bonds proved a challenge in my research into the materials at times – and also a sense of wonder at the times we live in, the relations that are beyond my ken. The kinds of bonds I have

consumerism, and, as I learnt from Doreen Massey a while back in conversation, we (here, teachers of teachers) might need to make spaces for unlearning these kinds of things, so that such thinking can multiply. An example: In a seminar, we (here: me and the students in the room) came across the Northwestern Pacific practice of potlatch in a reading. Some students (bravely!) indicated their lack of understanding for the wanton destruction of status symbols and wealth entailed by the practice. Thinking on my feet, I compared this practice to the possibly more recognizable practice of buying a new car: the idea being that as soon as you leave the car yard, the new car reduces in value by half (whether this is true or not doesn't matter, it was accepted as being mostly true and therefore the comparison works). The status symbol of buying a new car is also a wanton reduction of wealth. Can we be weirded out by that? Keller Easterling suggests that we might be able to employ design thinking to do this kind of tinkering (Easterling 2021).

[11] Gone: Teflon, Gore-Tex. No longer: reusing single-use plastic bottles. Phasing out: plastic boxes for food (I'm not throwing them out until they are finished, though). Some practices that look like thrift are also environmentalist.

come more familiar with are the ones that are forged by thinking through materialities: (imagined) relations.

Patricia Yaeger recalls a scene from Kurt Vonnegut Jr.'s *Slaughterhouse Five* in 'Dreaming of Infrastructure' (2007). Vonnegut Jr.'s protagonist, Billy, becomes 'slightly unstuck in time' (Vonnegut 1994: 70) and watches film footage of the bombing of Dresden in reverse. Formations of planes fly backwards over a city, exerting 'miraculous magnetism which shrunk the fires, gathered them into cylindrical steel containers, and lifted the containers into the bellies of the planes' (ibid.: 71), before returning to factories in the United States, which 'were operating day and night, dismantling the cylinders, separating the dangerous contents into minerals' (ibid.). The minerals are then 'shipped to specialists in remote areas' (ibid.), whose task was to 'hide them cleverly, so they would never hurt anybody ever again' (ibid.: 72). This is a passage that resonates with the temporal and spatial dimensions of materiality in the 'provenance story' mode, imagining – inventing – a place, the world, of 'no return'.

In Emily St John Mandel's *Station Eleven*, as in any number of post-apocalyptic or post-disaster scenarios, these relations are rendered present through absence. Their absence is felt keenly enough to be a gap, a break. Infrastructure, as Michael Rubenstein, Bruce Robbins and Sophia Beal suggest, like Yaeger above, 'is supposed to go unnoticed when it works' (Rubenstein, Robbins and Beal 2015: 576). Concrete works, in my experience, to shape my life, but is mostly unnoticed: concrete is mostly remarkably 'unconcrete' in this sense. Not just a 'sign taken for modernity', then, as much as also a 'sign *mis*taken for modernity' and a 'sign *un*taken for modernity'. This book affords these issues time and space. Thinking through concrete affords attention paid to the structure to the paths I take and to the paths of the facilities upon which I rely – buildings, roads, power supplies, water services and so on. It means grappling with the ways in which materials exert their materiality, beyond, without, beside human agency.

Thinking through materialities has also meant mobilizing interpretative practices *as* inventive (and interventive) practices. It means shifting critical practices away from the application of theoretical methods to otherwise (what? dormant?) texts. Instead, it draws on an understanding of interpretation as invention to foreground the 'coming into relations' that the latter suggests ('Introduction'). Part of this is considering generic concerns as aesthetic formations, not interpretative barriers: One kind of text does not act as a 'key' for 'unlocking' another only by virtue of its aesthetic or generic form. In the analyses, inventiveness is interpreted as a marker of 'coming into relations', rather than synonymous with an understanding of 'fictional' that posits an opposition to 'real'. Inventiveness pertains to a characteristic of imagination, of potential for (speculative) engagement that does not necessitate a – however constituted – separation from the 'feasible'. Artefacts, and/as texts, are inventive.

This inventiveness has specific consequences for the issues pertaining to scales outlined earlier. The 'provenance stories' are one of the forms that the familiar scales of materials can be reinvented, connecting disparate spaces along unfamiliar timelines with present presences. The many forms of the 'future artefact' – including the future archaeologist (see above), future fossils and the frameworks of waste – comprise another. Scales of the human – which comprise the inventor(s), the addressees, as well as the agents of most of these texts – can be flaunted or forgotten, as new (ways of understanding) the world(s) are forged.

By tracing and constituting imaginative entanglements with disparate others, textual analysis is a form of engagement in Anthropocenes, modernities and other (environmental, political) crises, an engagement in, and an entanglement with, material practices. Living with our leftovers, creating with castaways: thinking with materialities emerges as a set of practices encompassing inventory, invention *and* intervention.

Bibliography

Abbey, Edward. *The Monkey Wrench Gang*. New York: Harper Perennial, (1975) 2006.
Adamson, Joni. 'Indigenous Cosmopolitics and the Re-Emergence of the Pluriverse'. In *Howling for Justice: New Perspectives on Leslie Marmon Silko's Almanac of the Dead*, edited by Rebecca Tillett, 181-94. Tucson: University of Arizona Press, 2014.
Ahmed, Sara. 'Happy Objects'. In *The Affect Theory Reader*, edited by Melissa Gregg and Gregory J. Seigworth, 29-51. Durham, NC: Duke University Press, 2010.
Alaimo, Stacy. *Bodily Natures: Science, Environment, and the Material Self*. Bloomington: Indiana University Press, 2010.
Alaimo, Stacy. 'States of Suspension: Trans-corporeality at Sea'. *ISLE: Interdisciplinary Studies in Literature and Environment* 19, no. 3 (2012): 476-93.
Alaimo, Stacy. *Exposed: Environmental Politics and Pleasures in Posthuman Times*. Minneapolis: University of Minnesota Press, 2016.
Allen, Fiona. 'Introduction: Concrete'. *Parallax* 21, no. 3 (2015): 237-40.
Altinbek, Dogan. 'The Role of Dams in Development'. *International Journal of Water Resources Development* 18, no. 1 (2002): 9-24.
Anderson, Benedict. *Imagined Communities: Reflections on the Origin and Spread of Nationalism*, 2nd edn. London: Verso, 2006.
Appadurai, Arjun. 'Introduction: Commodities and the Politics of Value'. In *The Social Life of Things*, edited by Arjun Appadurai, 3-63. Cambridge: Cambridge University Press, 1986.
Armiero, Marco. *Wasteocene: Stories from the Global Dump*. Cambridge: Cambridge University Press, 2021.
Baal-Teshuva, Jacob. *Christo and Jeanne-Claude*. Cologne: Taschen, 2016.
Bailey, Rowan. 'Concrete Thinking for Sculpture'. *Parallax* 21, no. 3 (2015): 241-58.
Baldacchino, Godfrey, and Eric Clark. 'Guest Editorial Introduction: Islanding Cultural Geographies'. *Cultural Geographies* 20, no. 2 (2013): 129-34.
Ballard, J. G. *Concrete Island*. London: Fourth Estate, (1974) 2011.
Ballard, J. G. *High-Rise*. London: Fourth Estate, (1975) 2014.
Barad, Karen. *Meeting the Universe Halfway: Quantum Physics and the Entanglement of Matter and Meaning*. Durham, NC: Duke University Press, 2007.
Barber, Katrine E. 'Wisecracking Glen Canyon Dam: Revisioning Environmentalist Mythology'. In *Change in the American West: Exploring the Human Dimension*, edited by Stephen Tchudi, 127-43. Reno: University of Nevada Press, 1996.
Barthes, Roland. *Mythologies*, trans. Annette Lavers. London: Vintage, 2000.

Bassnett, Susan, and Harish Trivedi. 'Introduction: Of Colonies, Cannibals and Vernaculars'. In *Post-Colonial Translation: Theory and Practise*, edited by Susan Bassnett and Harish Trivedi, 1–18. London: Routledge, 2002.

Beck, John. 'Concrete Ambivalence: Inside the Bunker Complex'. *Cultural Politics* 7, no. 1 (2011): 79–102.

Bellis, Mary. 'Who Invented Saran Wrap?', thoughtco.com, 1 September 2017. Available online: https://www.thoughtco.com/history-of-pvdc-4070927 (accessed 28 June 2018).

Bennett, Jane. *Vibrant Matter: A Political Ecology of Things*. Durham, NC: Duke University Press, 2010.

Bennett, Luke. 'Concrete Multivalence: Practising Representation in Bunkerology'. *Environment and Planning D: Society and Space* 31, no. 3 (2013): 502–21.

Bhabha, Homi. 'Signs Taken for Wonders: Questions of Ambivalence and Authority under a Tree outside Delhi, May 1817'. *Critical Inquiry* 12, no. 1 (1985): 144–65.

Bhambra, Gurminder K. *Rethinking Modernity: Postcolonialism and the Sociological Imagination*. Basingstoke: Palgrave Macmillan, 2007.

Biamonti, Alessandro. *Archiflop: A Guide to the Most Spectacular Failures in the History of Modern and Contemporary Architecture*, trans. Susan Ann White, Felicity Lutz and Lauren Sunstein. Zurich: Niggli, 2017.

Blackwell, Andrew. *Visit Sunny Chernobyl: Adventures in the World's Most Polluted Places*. New York: Arrow Books, 2013.

Blezard, Robert G. 'The History of Calcareous Cements'. In *Lea's Chemistry of Cement and Concrete*, edited by Peter C. Hewlett, 1–23, 4th edn. Amsterdam: Elsevier, 2004.

Braun, Wilhelm, and Wolfgang Pfeifer. *Etymologisches Wörterbuch des Deutschen*, volume 1 (A–G). Berlin: Akademie Verlag, 1989.

Brinkley, Douglas. 'Introduction'. In *The Monkey Wrench Gang*, by Edward Abbey, xv–xxiv. New York: Harper Perennial, 2006.

Brosch, Renate, and Kylie Crane. 'Visualising Australia: An Introduction'. In *Visualising Australia: Images, Icons, Imaginations*, edited by Renate Brosch and Kylie Crane, 1–17. Trier: WVT, 2014.

Brown, Bill. 'Thing Theory'. *Critical Inquiry* 28, no. 1 (2001): 1–22.

Brown, Kate. 'Learning to Read the Great Chernobyl Acceleration: Literacy in the More-than-Human Landscapes'. *Current Anthropology* 60, no. S20 (2019): 198–209.

Bunn, Stephanie. 'The Importance of Materials'. *Journal of Museum Ethnography* 11 (1999): 15–28.

Carle, Eric. *10 Little Rubber Ducks*. New York: HarperFestival, 2010.

Carruthers, Mary. *The Craft of Thought: Meditation, Rhetoric, and the Making of Images 400–1200*. Cambridge: Cambridge University Press, 2000.

Carson, Rachel. *Silent Spring*. New York: Fawcett Crest, 1962.

Carter, Erica, James Donald and Judith Squires. 'Introduction'. In *Space and Place: Theories of Identity and Location*, edited by Erica Carter, James Donald and Judith Squires, vii–xv. London: Lawrence and Wishart, 1993.

Carter, Paul. 'Incontinence: Australia and the Archipelago'. In *Visualising Australia: Images, Icons, Imaginations*, edited by Renate Brosch and Kylie Crane, 181–97. Trier: WVT, 2014.

Chakrabarty, Dipesh. 'The Climate of History: Four Theses'. *Critical Inquiry* 35, no. 2 (2009): 197–22.

Chantrell, Glynnis, ed. *The Oxford Dictionary of Word Histories*. Oxford: Oxford University Press, 2002.

Cho, Renee. 'The Truth About Bioplastics', *State of the Planet*. Earth Institute, Columbia University, 13 December 2017. Available online: https://blogs.ei.columbia.edu/2017/12/13/the-truth-about-bioplastics/ (accessed 21 November 2018).

Clavé, Salvador Anton, Filippo Carlà-Uhink and Florian Freitag. 'Time: Represented, Experienced, and Managed Temporalities in Theme Parks'. In *Key Concepts in Theme Park Studies: Understanding Tourism and Leisure Spaces*, coordinated by Florian Freitag, Filippo Carlà-Uhink, and Salvador Anton Clavé, 309–22. Cham: Springer, 2023.

Cocker, Mark. 'Death of the Naturalist: Why Is the "New Nature Writing" So Tame?', *The New Statesman*, 17 June 2015. Available online: http://www.newstatesman.com/culture/2015/06/death-naturalist-why-new-nature-writing-so-tame (accessed 11 January 2016).

Colebrook, Claire. *Death of the PostHuman: Essays on Extinction, Vol. 1*. Ann Arbor: Open Humanities Press, 2014.

'Concrete'. In *Concise Oxford English Dictionary*, edited by Angus Stevenson and Maurice Waite, 299, 12th edn. Oxford: Oxford University Press, 2011.

Coole, Diana, and Samantha Frost. *New Materialisms: Ontology, Agency, and Politics*. Durham, NC: Duke University Press, 2010a.

Coole, Diana, and Samantha Frost. 'Introducing the New Materialisms'. In *New Materialisms: Ontology, Agency, and Politics*, edited by Diana Coole and Samantha Frost, 1–46. Durham, NC: Duke University Press, 2010b.

Corcoran, Patricia L., Charles J. Moore and Kelly Jazvac. 'An Anthropogenic Marker Horizon in the Future Rock Record'. *GSA Today* 24, no. 6 (June 2014): 4–8. Available online: http://www.geosociety.org/gsatoday/archive/24/6/pdf/i1052-5173-24-6-4.pdf (accessed 24 March 2015).

Cox, Kieran D., Garth Covernton, Hailey Davies, John F. Dower, Francis Juanes and Sarah E. Dudas. 'Human Consumption of Microplastics'. *Environmental Science & Technology* 52, no. 12 (2019): 7068–74.

Crane, Kylie. 'Plastic Modernities'. In *Anglistentag 2015 Paderborn*, edited by Christoph Ehland, Ilka Mindt and Merle Tönnies, 207–17. Trier: WVT, 2016.

Crane, Kylie. 'Ecocriticism and Travel'. In *Cambridge History of Travel Writing*, edited by Nandini Das and Tim Youngs, 535–49. Cambridge: Cambridge University Press, 2019a.

Crane, Kylie. 'Anthropocene Presences and the Limits of Deferral: Alexis Wright's *Carpentaria* and *The Swan Book*'. *Online Library of Humanities* 5, no. 1 (2019b), edited by Timo Müller, special issue: 'Representing Climate: Local to Global'. doi: 10.16995/olh.348.

Crane, Kylie. 'Thinking Fungi, or Random Considerations'. *Comparative Critical Studies* 18, nos. 2–3 (2021a): 239–58. doi: 10.3366/ccs.2021.0405.

Crane, Kylie. 'Displacements: Framing (and) Ruins in John Berger's King and Indra Sinha's Animal's People'. *Open Library of Humanities* 7 no. 1, 2021b. doi: 10.16995/Olh.633.

Crane, Kylie. 'Birds of the Plastic Pacific'. In *Maritime Mobilities: Literary and Cultural Perspectives from the Anglophone World*, edited by Alexandra Ganser and Charne Lavery, 41–56. New York: Palgrave Macmillan, 2023a.

Crane, Kylie. 'On Some Absent Presences of Nuclear Extractivism: Retrofuturist Aesthetics and *Fallout 4*'. In *To the Last Drop – Affective Economies of Extraction and Sentimentality*, edited by Axelle Germanaz, Daniela Gutiérrez Fuentes, Sarah Marak and Heike Paul, 185–202. Bielefeld: Transcript, 2023b.

Crane, Kylie. 'Robinsonades Revisited: Materials and Archives in the Archipelagic Mode'. *PhiN. Philologie im Netz: Beihefte* 30 (2023c): 7–20. https://web.fu-berlin.de/phin/beiheft30/b30t2.pdf (special issue *(Re-)Framing the Robinsonade*).

Cresswell, Tim. *Plastiglomerate*. London: Penned in the Margins, 2020.

Culler, Jonathan. *Framing the Sign: Criticism and Its Institutions*. Oxford: Blackwell, 1988.

Curry, Andrew. 'With Bernhard Ludewig (Photography). "What Lies Beneath"'. *Atlantic* 320, no. 2 (2017): 52–7.

Davis, Heather. 'Life & Death in the Anthropocene: A Short History of Plastic'. In *Art in the Anthropocene: Encounters Among Aesthetics, Politics, Environments and Epistemologies*, edited by Heather Davis and Etienne Turpin, 347–58. London: Open Humanities Press, 2015.

Davis, Heather. *Plastic Matter*. Durham, NC: Duke University Press, 2022.

Davis, Janae, Alex A. Moulton, Levi Van Sant and Brian Williams. 'Anthropocene, Capitalocene, … Plantationocene?: A Manifesto for Ecological Justice in an Age of Global Crises'. *Geography Compass* 13, no. 5 (2019): e12438. doi: 10.1111/gec3.12438.

de Botton, Alain. *The Pleasures and Sorrows of Work*. London: Hamish Hamilton, 2009.

Deckard, Sharae. '"Inherit the World": World-Literature, Rising Asia, and the World-Ecology'. In *What Postcolonial Theory Doesn't Say*, edited by Bernard, Anna, Ziad Elmarsafy and Stuart Murray, 239–56. London: Routledge, 2015.

DeLoughrey, Elizabeth. '"The Litany of Islands, The Rosary of Archipelagos"': Caribbean and Pacific Archipelagraphy'. *ARIEL: A Review of International English Literature* 32, no. 1 (2001): 21–51.

DeLoughrey, Elizabeth. *Routes and Roots: Navigating Caribbean and Pacific Island Literatures*. Honolulu: University of Hawai'i Press, 2007.

DeLoughrey, Elizabeth. 'The Myth of Isolates: Ecosystem Ecologies in the Nuclear Pacific'. *Cultural Geographies* 20, no. 2 (2012): 167–84.

Deyo, L. B., and David 'Lefty' Leibowitz. *Invisible Frontier: Exploring the Tunnels, Ruins, and Rooftops of Hidden New York*. New York: Three Rivers Press, 2003.

Doss, Erika. 'Public Art, Public Feeling: Contrasting Site-Specific Projects of Christo and Ai Weiwei'. *Public Art Dialogue* 7, no. 2 (2017): 196–229.

Douglas, Mary. *Purity and Danger: An Analysis of Concepts of Pollution and Taboo*. London: Routledge, (1966) 2002.

Easterling, Keller. *Medium Design: Knowing How to Work on the World*. London: Verso, 2021.

Ebbesmeyer, Curtis, and Eric Scigliano. *Flotasmetrics and the Floating World*. New York: HarperCollins, 2009.

Eco, Umberto. *Travels in Hyperreality*, trans. William Weaver. London: Picador, 1987.

Eckstein, Lars, and Anja Schwarz. 'Introduction: Towards a Postcolonial Critique of Modern Piracy'. In *Postcolonial Piracy: Media Distribution and Cultural Production in the Global South*, edited by Eckstein, Lars and Anja Schwarz, 1–25. London: Bloomsbury, 2014.

Edensor, Tim. 'Waste Matter: The Debris of Industrial Ruins and the Disordering of the Material World'. *Journal of Material Culture* 10, no. 3 (2005a): 311–32.

Edensor, Tim. *Industrial Ruins: Spaces, Aesthetics and Materiality*. Oxford: Berg, 2005b.

Eichhoff, Jürgen. 'Unterschiede des Wortgebrauchs'. In *Sprache und Brauchtum: Bernhard Martin zum 90. Geburtstag*, edited by Reiner Hildebrandt and Hans Freibertshäser, 154–78. Marburg: N. G. Elwert Verlag, 1980.

Epstein, Rebecca, Jamilia J. Blake and Thalia González. 'Girlhood Interrupted: The Erasure of Black Girls' Childhood', *Center on Poverty and Inequality (Georgetown Law)*, 27 June 2017. Available online: https://ssrn.com/abstract=3000695 (accessed 17 June 2018).

Escobar, Arturo. *Encountering Development: The Making and Unmaking of the Third World*. Princeton, NJ: Princeton University Press, 1995.

Fahim, Hussein M. *Dams, People and Development: The Aswan High Dam Case*. New York: Pergamon Press, 1981.

Farley, Paul, and Michael Symmons Roberts. *Edgelands: Journeys into England's True Wilderness*. London: Vintage, (2011) 2012.

Farmer, Jared. *Glen Canyon Dammed: Inventing Lake Powell and the Canyon Country*. Tucson: University of Arizona Press, 1999.

Farmer, Jared. 'Future Fossil'. The Anthropocene Slam: A Cabinet of Curiosities, 9 November 2014. Available online: http://nelson.wisc.edu/che/anthroslam/objects/farmer.php (accessed 14 September 2018).

Farmer, Jared. 'Technofossil'. In *Future Remains: A Cabinet of Curiosities for the Anthropocene*, edited by Gregg Mitman, Marco Armiero and Robert S. Emmett, 191–9. Chicago: University of Chicago Press, 2018.

Farrier, David. *Footprints: In Search of Future Fossils*. London: Fourth Estate, 2020.

Fenichell, Stephen. *Plastic: The Making of a Synthetic Century*. New York: HarperCollins, 1996.

Fisher, Tom. 'The Death and Life of Plastic Surfaces: Mobile Phones'. In *Accumulation: The Material Politics of Plastic*, edited by Gay Hawkins, Jennifer Gabrys and Mike Michael, 107–20. London: Routledge, 2013.

'Fluid'. In *Concise Oxford English Dictionary*, edited by Angus Stevenson and Maurice Waite, 548, 12th edn. Oxford: Oxford University Press, 2011.

Foreman, David. *Ecodefense: A Field Guide to Monkeywrenching*. Chico, CA: Abbzug Press, 1993.

Forty, Adrian. *Concrete and Culture: A Material History*. London: Reaktion Books, 2012.

Foucault, Michel. *The Archaeology of Knowledge & The Discourse on Language*, trans. Alan Sheridan. 1969. New York: Pantheon Books, 1972.

Freinkel, Susan. *Plastic: A Toxic Love Story*. Boston: Houghton Mifflin Harcourt, 2011.

Gabrys, Jennifer. 'Plastic and the Work of the Biodegradable'. In *Accumulation: The Material Politics of Plastic*, edited by Gay Hawkins, Jennifer Gabrys and Mike Michael, 208–27. London: Routledge, 2013.

Gaonkar, Dilip Parameshwar. 'On Alternative Modernities'. *Public Culture* 11, no. 1 (1999): 1–18.

Garrard, Greg. 'Ecocriticism'. *The Year's Work in Critical and Cultural Theory* 20, no. 1 (2012): 200–43.

Garrett, Bradley L. *Explore Everything: Place-Hacking the City*. London: Verso, (2013) 2014.

Groes, Sebastian. 'The Texture of Modernity in J. G. Ballard's *Crash*, *Concrete Island* and *High-Rise*'. In *J.G. Ballard: Visions and Revisions*, edited by Jeanette Baxter and Rowland Wymer, 123–41. Basingstoke: Palgrave Macmillan, 2012.

Gunesekera, Romesh. *Reef*. London: Granta, 1994.

Hall, Stuart. 'Cultural Identity and Diaspora'. In *Identity. Community, Culture, Difference*, edited by Jonathan Rutherford, 222–37. London: Lawrence and Wishart, 1990.

Hall, William, ed. *Concrete*. London: Phaidon Press, 2012.

Haller, John S., Jr. *Outcasts from Evolution: Scientific Attitudes of Racial Inferiority 1859–1900*. Carbondale: Southern Illinois University Press, 1995.

Hamid, Mohsin. *How to Get Filthy Rich in Rising Asia*. London: Hamish Hamilton, 2013.

Hamid, Mohsin. *Exit West*. London: Penguin, 2017.

Hansen, Gary. *Wet Desert*. USA: Hole Shot Press, 2007.

Haraway, Donna. 'Situated Knowledges: The Science Question in Feminism and the Privilege of Partial Perspective'. *Feminist Studies* 14, no. 3 (1988): 575–99.

Haraway, Donna. *The Companion Species Manifesto: Dogs, People, and Significant Otherness*. Chicago: Prickly Paradigm Press, 2003.

Haraway, Donna. *Staying with the Trouble: Making Kin in the Chthulucene*. Durham, NC: Duke University Press, 2016.

Harkness, Rachel, Cristián Simonetti and Judith Winter. 'Liquid Rock'. *The Anthropocene Slam: A Cabinet of Curiosities*, 10 November 2014. Available online: http://nelson.wisc.edu/che/anthroslam/objects/simon.php (accessed 14 September 2018).

Harkness, Rachel, Cristián Simonetti and Judith Winter. 'Liquid Rock: Gathering, Flattening, Curing'. *Parallax* 21, no. 3 (2015): 309–25.

Harkness, Rachel, Cristián Simonetti and Judith Winter. 'Concretes Speak'. In *Future Remains: A Cabinet of Curiosities for the Anthropocene*, edited by Gregg Mitman, Marco Armiero and Robert S. Emmett, 29–39. Chicago: University of Chicago Press, 2018.

Hashimoto, Isao. 'Time-Lapse Map of All 2053 Nuclear Explosions 1945–1998', youtube.com. Available online: https://www.youtube.com/watch?v=dxyRLvcjVCw (accessed 20 April 2018).

Hawkins, Gay. 'Packaging Water: Plastic Bottles as Market and Public Devices'. *Economy and Society* 40, no. 4 (2011): 534–52.

Hawkins, Gay. 'Made to be Wasted: PET and Topologies of Disposability'. In *Accumulation: The Material Politics of Plastic*, edited by Gay Hawkins, Jennifer Gabrys and Mike Michael, 49–67. London: Routledge, 2013.

Hawkins, Gay, Jennifer Gabrys and Mike Michael, eds. *Accumulation: The Material Politics of Plastic*. London: Routledge, 2013.

Hawkins, Gay, Emily Potter and Kane Race. *Plastic Water: The Social and Material Life of Bottled Water*. Cambridge: Massachusetts Institute of Technology Press, 2015.

Hecht, Gabrielle. 'Interscalar Vehicles for an African Anthropocene: On Waste, Temporality, and Violence'. *Cultural Anthropology* 33, no. 1 (2018): 109–41.

Heise, Ursula K. 'Toxins, Drugs, and Global Systems: Risk and Narrative in the Contemporary Novel'. *American Literature* 74, no. 4 (2002): 747–78.

Hickman, Bill. 'Fly or Die – with Chris Jordan'. Surfrider.org, 26 June 2012. Available online: http://www.surfrider.org/coastal-blog/entry/chris-jordan-interview (accessed 2 November 2015).

Hodder, Ian. *Entangled: An Archaeology of the Relationships between Humans and Things*. Malden: John Wiley, 2012.

Hogan, Linda. *Solar Storms*. New York: Scribner, 1997.

Hohn, Donovan. *Moby-Duck: The True Story of 28,800 Bath Toys Lost at Sea*. London: Union Books, 2012.

Howe, Cymene, Jessica Lockrem, Hanna Appel, Edward Hackett, Dominic Boyer, Randal Hall, Matthew Schneider-Mayerson, Albert Pope, Akhil Gupta, Elizabeth Rodwell, Andrea Ballestero, Trevor Durbin, Farès el-Dahdah, Elizabeth Long and Cyrus Mody. 'Paradoxical Infrastructures: Ruins, Retrofit, and Risk'. *Science, Technology, & Human Values* 41, no. 3 (2016): 547–65.

Ialenti, Vincent F. 'Adjudicating Deep Time: Revisiting the United States' High-Level Nuclear Waste Repository Project at Yucca Mountain'. *Science & Technology Studies* 27, no. 2 (2014): 27–48.

Ingold, Tim. 'Materials against Materiality'. *Archaeological Dialogues* 14, no. 1 (2007): 1–16.

Iovino, Serenella, and Serpil Oppermann. 'Theorizing Material Ecocriticism: A Diptych'. *ISLE: Interdisciplinary Studies in Literature and Environment* 19, no. 3 (2012): 448–75.

Iovino, Serenella, and Serpil Oppermann. *Material Ecocriticism*. Bloomington: Indiana University Press, 2014a.

Iovino, Serenella, and Serpil Oppermann. 'Introduction: Stories Come to Matter'. In *Material Ecocriticism*, edited by Iovino, Serenella and Serpil Oppermann, 1–17. Bloomington: Indiana University Press, 2014b.

Jackson, Mark. 'Plastic Islands and Processual Grounds: Ethics, Ontology and the Matter of Decay'. *Cultural Geographies* 20, no. 2 (2012): 205–24.

Jetñil-Kijiner, Kathy. 'Anointed'. April 2018. Available as video online produced by PREL, directed by Dan Lin, https://www.youtube.com/watch?v=HuDA7izeYrk (accessed 27 March 2023).

Jones, Alison, and Te Kawehau Hoskins. 'A Mark on Paper: The Matter of Indigenous – Settler History'. In *Posthuman Research Practices in Education*, edited by Carol A. Taylor and Christina Hughes, 75–92. Basingstoke: Palgrave Macmillan, 2016.

Jordan, Chris. 'About', February 2011. Available online: http://www.chrisjordan.com/gallery/midway/#about (accessed 28 April 2015).

Kakutani, Michiko. 'Love and Ambition in a Cruel New World: *How to Get Filthy Rich in Rising Asia* by Mohsin Hamid', nytimes.com, 21 February 2013. Available online: https://www.nytimes.com/2013/02/22/books/how-to-get-filthy-rich-in-rising-asia-by-mohsin-hamid.html (accessed 28 June 2018).

Kenny, John J., and Ronald R. Hrusoff. 'The Ownership of the Treasures of the Sea'. *William & Mary Law Review* 383 (1967): 383–401. Available online: http://scholarship.law.wm.edu/wmlr/vol9/iss2/7 (accessed February 2013).

King, Thomas. *Green Grass, Running Water*. New York: Bantam, (1993) 1994.

King, Thomas. *The Inconvenient Indian: A Curious Account of Native People in North America*. Minneapolis: University of Minnesota Press, 2012.

Klein, Naomi. *This Changes Everything: Climate vs Capitalism*. New York: Simon & Schuster, 2014.

Koren, Leonard. 'Concrete Thoughts'. In *Concrete*, edited by William Hall, 8–13. London: Phaidon Press, 2012.

Laferrière, Dany. *The World is Moving around Me: A Memoir of the Haiti Earthquake*, trans. David Homel. Vancouver: Arsenal Pulp Press, (2011) 2013.

Lam, Tong. *Abandoned Futures*. Darlington: Carpet Bombing Culture, 2013.

Larkin, Brian. 'The Politics and Poetics of Infrastructure'. *Annual Review of Anthropology* 42 (2013): 327–43.

Latour, Bruno. *We Have Never Been Modern*, trans. Catherine Porter. Cambridge: Harvard University Press, 1993.

Lavery, Charne. 'Indian Ocean Depths: Cables, Cucumbers, Consortiums'. *The Johannesburg Salon* 10 (2015): 26–9. Once available online: http://jwtc.org.za/resources/docs/salon-volume-10/JWTC_vol_10_salon.pdf (accessed 21 May 2018).

Le Corbusier (Jeanneret, Charles-Édouard). *Towards an Architecture*, trans. Frederick Etchells. New York: Dover, (1931) 1986.
LeMenager, Stephanie. *Living Oil: Petroleum Culture in the American Century*. Oxford: Oxford University Press, 2014.
Lepkowski, Wil. 'The Restructuring of Union Carbide'. In *Learning from Disaster: Risk Management after Bhopal*, edited by Sheila Jasanoff, 22–43. Philadelphia: University of Pennsylvania Press, 1994.
Lewis, Simon L., and Maslin, Mark A. 'Defining the Anthropocene'. *Nature* 519, no. 12 (2015): 171–80.
Liboiron, Max. *Pollution is Colonialism*. Durham, NC: Duke University Press, 2021.
Liboiron, Max, Manuel Tironi and Nerea Calvillo. 'Toxic Politics: Acting in a Permanently Polluted World'. *Social Studies of Science* 48, no. 3 (2018): 331–49.
Liittschwager, David, and Susan Middleton. *Archipelago: Portraits of Life in the World's Most Remote Island Sanctuary*. Washington, DC: National Geographic Society, 2005.
Lindholt, Paul. *Explorations in Ecocriticism: Advocacy, Bioregionalism, and Visual Design*. Boston: Lexington Press, 2015.
Lipinski, Klaus. 'Spurensammler', *Lipinski.de* (n.d.). Available online: http://www.lipinski.de/ (accessed 27 June 2019).
'Liquid'. In *Concise Oxford English Dictionary*, edited by Angus Stevenson and Maurice Waite, 830, 12th edn. Oxford: Oxford University Press, 2011.
Lloyd Thomas, Katie. 'Rendered Plastic by Preparation: Concrete as Constant Material'. *Parallax* 21, no. 3 (2015): 271–87.
Luckhurst, Roger. *'The Angle between Two Walls': The Fiction of J.G. Ballard*. Liverpool: Liverpool University Press, 1997.
Lyng, Stephen. 'Edgework: A Social Psychological Analysis of Voluntary Risk Taking'. *American Journal of Sociology* 95, no. 4 (1990): 851–86.
Macfarlane, Robert. 'Why We Need Nature Writing', *New Statesman*, 2 September 2015. Available online: http://www.newstatesman.com/culture/nature/2015/09/robert-macfarlane-why-we-need-nature-writing (accessed 11 January 2016).
Macfarlane, Robert. *Underland: A Deep Time Journey*. London: Hamish Hamilton, 2019.
Mandel, Emily St. John. *Station Eleven*. New York: Picador, 2014.
Marriott, James, and Mika Minio-Paluello. 'Where Does This Stuff Come From? Oil, Plastic and the Distribution of Violence'. In *Accumulation: The Material Politics of Plastic*, edited by Gay Hawkins, Jennifer Gabrys and Mike Michael, 171–83. London: Routledge, 2013.
Marshall, George. *Don't Even Think About It: Why Our Brains Are Wired to Ignore Climate Change*. London: Bloomsbury, 2015.
Masco, Joseph. *The Future of Fallout, and Other Episodes in Radioactive World-Making*. Durham, NC: Duke University Press, 2021.
Massey, Doreen. *For Space*. London: Sage, 2006.
Matsuda, Matt K. *Pacific Worlds: A History of Seas, Peoples, and Cultures*. Cambridge: Cambridge University Press, 2016.

Matza, Tomas, and Nicole Heller. 'Anthropocene in a Jar'. In *Future Remains: A Cabinet of Curiosities for the Anthropocene*, edited by Gregg Mitman, Marco Armiero and Robert S. Emmett, 22–8. Chicago: University of Chicago Press, 2018.

McCann, Joy. *Balancing Act: The Australian Greens 2008–2011*. Department of Parliamentary Services, Parliamentary Library, 2012. Available online: https://australianpolitics.com/downloads/greens/2012/12-02-08_balancing-act-the-aust-greens-2008-11_mccann_aph.pdf (accessed 27 March 2023).

McCully, Patrick. *Silenced Rivers: The Ecology and Politics of Large Dams*. London: Zed Books, 2001.

Meikle, Jeffrey L. *American Plastic: A Cultural History*. New Brunswick: Rutgers University Press, 1995.

Michaels, Anne. *The Winter Vault*. London: Bloomsbury, (2009) 2010.

Mildorf, Jarmila. 'Pragmatic Implications of "You"-Narration for Postcolonial Fiction: Mohsin Hamid's *How to Get Filthy Rich in Rising Asia*'. In *Pragmatic Perspectives on Postcolonial Discourse: Linguistics and Literature*, edited by Christoph Schubert and Laurenz Volkmann, 99–113. Newcastle upon Tyne: Cambridge Scholars Publishing, 2016.

'Milestones'. *UNESCO (United Nations Educational, Scientific and Cultural Organisation)*. Once available online: http://www.unesco.org/new/en/unesco/about-us/who-we-are/history/milestones/ (accessed 7 June 2018).

Miodownik, Mark. *Stuff Matters: The Strange Stories of the Marvellous Materials That Shape Our Man-made World*. London: Penguin, 2014.

Mitchell, Audra. 'Decolonising the Anthropocene', worldlyir.wordpress.com, 17 March 2015. Available online: https://worldlyir.wordpress.com/2015/03/17/decolonising-the-anthropocene/ (accessed 26 October 2018).

Mitman, Gregg, Marco Armiero and Robert S. Emmett, eds. *Future Remains: A Cabinet of Curiosities for the Anthropocene*. Chicago: University of Chicago Press, 2018.

Molotch, Harvey Luskin. *Where Stuff Comes From: How Toasters, Toilets, Cars, Computers, and Many Other Things Came to Be as They Are*. New York: Routledge, 2003.

Moore, Charles, and Cassandra Phillips. *Plastic Ocean: How a Sea Captain's Chance Discovery Launched a Determined Quest to Save the Oceans*. New York: Avery, 2012.

Moore, Jason. 'The Capitalocene Part I: On the Nature and Origins of our Ecological Crisis'. *Journal of Peasant Studies* 44, no. 3 (2017): 594–630.

Moore, Jason. 'The Capitalocene Part II: Accumulation by Appropriation and the Centrality of Unpaid Work'. *Journal of Peasant Studies* 45, no. 2 (2018): 237–79.

More, Thomas. *Utopia*, trans. Paul Turner. London: Penguin, 1965.

Morton, Timothy. *The Ecological Thought*. Cambridge: Harvard University Press, 2010.

Motlagh, Jason. 'The Ghosts of Rana Plaza', *VQR*, Spring 2014. Available online: https://www.vqronline.org/reporting-articles/2014/04/ghosts-rana-plaza (accessed 14 June 2018).

Murphy, Patrick D. 'Damning Damming Modernity: The Destructive Role of Megadams'. *Tamkang Review* 42, no. 1 (2011): 27–40.
Nathan, Emily. 'Arnet Asks: Epic Environmental Artist Christo', Artnet.com, 29 October 2015. Available online: https://news.artnet.com/exhibitions/artnet-asks-christo-350807 (accessed 28 June 2018).
'Nexus'. In *Concise Oxford English Dictionary*, edited by Angus Stevenson and Maurice Waite, 965, 12th edn. Oxford: Oxford University Press, 2011.
Nielsen, Tom. 'The Return of the Excessive: Superfluous Landscapes'. *Space and Culture* 5, no. 1 (2002): 53–62.
Ninjalicious. *Access All Areas: A User's Guide to the Art of Urban Exploration*. Canada: Infilpress, 2005.
Nixon, Rob. 'Unimagined Communities: Developmental Refugees, Megadams and Monumental Modernity'. *New Formations* 69 (2010): 62–80.
Nixon, Rob. *Slow Violence and the Environmentalism of the Poor*. Cambridge, MA: Harvard University Press, 2011.
Nixon, Robert. 'The Anthropocene: The Promise and Pitfalls of an Epochal Idea'. In *Future Remains: A Cabinet of Curiosities for the Anthropocene*, edited by Gregg Mitman, Marco Armiero and Robert S. Emmett, 1–18. Chicago: University of Chicago Press, 2018.
Oliveira, Gil, Eric Dorfman, Nicolas Kramar and Chase Mendenhall. 'The Anthropocene in Natural History Museums: A Productive Lens of Engagement'. *Curator: The Museum Journal* 63, no. 3 (2020): 333–51.
Parikka, Jussi. *The Anthrobscene*. Minneapolis: University of Minnesota Press, 2014.
Parikka, Jussi. *A Geology of Media*. Minneapolis: University of Minnesota Press, 2015.
Patel, Raj, and Jason. W. Moore. *A History of the World in Seven Cheap Things: A Guide to Capitalism, Nature, and the Future of the Planet*. London: Verso, 2020.
Penny, Louise. *How The Light Gets In*. London: Hodder & Staughton, (2013) 2021.
Perez, Craig Santos. 'Transterritorial Currents and the Imperial Terripelago'. *American Quarterly* 67, no. 3 (2015): 619–24.
Perez, Craig Santos. 'Guam and Archipelagic American Studies'. In *Archipelagic American Studies*, edited by Brian Russell Roberts and Michelle Ann Stephens, 97–112. Durham, NC: Duke University Press, 2017.
Perez, Craig Santos. *Habitat Threshold*. Oakland: Omnidawn Publishing, 2020.
Perez, Craig Santos. 'Thinking (and Feeling) with Anthropocene (Pacific) Islands'. *Dialogues in Human Geography* 11, no. 3 (2021): 429–33. doi: 10.1177/20438206211017453.
Pétursdóttir, Þóra. 'Concrete Matters: Ruins of Modernity and the Things called Heritage'. *Journal of Social Archaeology* 13, no. 1 (2013): 31–53.
Pickering, Andrew. 'Material Culture and the Dance of Agency'. In *The Oxford Handbook of Material Culture Studies*, edited by Dan Hicks and Mary C. Beaudry, 191–208. Oxford: Oxford University Press, 2010.

Pinder, David. 'Arts of Urban Exploration'. *Cultural Geographies* 12 (2005): 383–411.
'Plastic'. In *Concise Oxford English Dictionary*, edited by Angus Stevenson and Maurice Waite, 1098, 12th edn. Oxford: Oxford University Press, 2011.
PlasticsEurope (PEMRG). 'Global Plastic Production from 1950 to 2016 (in Million Metric Tons)'. Statista – The Statistics Portal, 2019. Available online: http://www.statista.com/statistics/282732/global-production-of-plastics-since-1950/ (accessed 26 June 2019).
Plumwood, Val. *Feminism and the Mastery of Nature*. London: Routledge, 1993.
Poon, Angelia. 'Helping the Novel: Neoliberalism, Self-Help, and the Narrating of the Self in Mohsin Hamid's *How to Get Filthy Rich in Rising Asia*'. *Journal of Commonwealth Literature* 52, no. 1 (2015): 139–50.
Pratt, Mary Louise. *Imperial Eyes: Travel Writing and Transculturation*. 1992. London: Routledge, 2003.
Premoli, Martin. '"We are fighting": Global Indigeneity and Climate Change'. *Transmotion* 7, no. 2 (2021): 1–26. doi: 10.22024/UniKent/03/tm.1041.
Probyn, Elspeth. 'Swimming with Tuna'. *Australian Humanities Review* 51 (2011). Available online: http://australianhumanitiesreview.org/2011/11/01/swimming-with-tuna-human-ocean-entanglements/ (accessed 28 June 2018).
Rathje, William, and Cullen Murphy. *Rubbish! The Archaeology of Garbage*. Tucson: University of Arizona Press, 2001.
Richardson, Brian. 'Keeping You Unnatural: Against the Homogenization of Second Person Writing. A Response to Joshua Parker'. *Connotations* 23, no. 1 (2013/2014): 49–54.
Roberts, Brian Russell, and Michelle Ann Stephens. 'Archipelagic American Studies and the Caribbean'. *Journal of Transnational American Studies* 5, no. 1 (2013). Available online: https://escholarship.org/uc/item/52f2966r (accessed 3 February 2018).
Roberts, Jody A. 'Reflections of an Unrepentant Plastiphobe: An Essay on Plasticity and the STS Life'. In *Accumulation: The Material Politics of Plastic*, edited by Gay Hawkins, Jennifer Gabrys and Mike Michael, 121–33. London: Routledge, 2013.
Ross, Derek G. 'Monkeywrenching Plain Language: Ecodefense, Ethics, and the Technical Communication of Ecotage'. *IEEE Transactions on Professional Communication* 58, no. 2 (2015): 154–75.
Roy, Arundhati. *An Ordinary Person's Guide to Empire*. New Delhi: Penguin, 2006.
Roy, Arundhati. *Broken Republic*. New Delhi: Hamish Hamilton, 2011.
Roy, Arundhati. *Listening to Grasshoppers: Field Notes on Democracy*. New Delhi: Penguin, (2009) 2013.
Roy, Arundhati. *Capitalism: A Ghost Story*. Chicago: Haymarket Books, 2014.
Rubenstein, Madeleine. 'Emissions from the Cement Industry'. *State of the Planet*, 9 May 2012. Available online: https://blogs.ei.columbia.edu/2012/05/09/emissions-from-the-cement-industry/ (accessed 27 June 2019).
Rubenstein, Michael, Bruce Robbins and Sophia Beal. 'Infrastructuralism: An Introduction'. *Modern Fiction Studies* 61, no. 4 (2015): 575–86.

Sedgwick, Eve Kosofsky. *Touching Feeling: Affect, Pedagogy, Performativity.* Durham, NC: Duke University Press, 2003.

Seymour, Nicole. *Bad Environmentalism: Irony and Irreverence in the Ecological Age.* Minneapolis: University of Minnesota Press, 2018.

Shackleford, Laura. 'Counter-Networks in a Network Society: Leslie Marmon Silko's *Almanac of the Dead*'. *Postmodern Culture* 16, no. 3, May 2006. Available online: http://pmc.iath.virginia.edu/issue.506/16.3shackelford.html (accessed 7 June 2018).

Shakespeare, William. *Hamlet.* Edited by Harold Jenkins. London: Arden, (1603) 2001.

Sheller, Mimi. *Aluminum Dreams: The Making of Light Modernity.* Cambridge, MA: Massachusetts Institute of Technology Press, 2014.

Shelley, Percy Bysshe. 'Ozymandias'. In *Shelley: Poetical Works*, edited by Thomas Hutchinson, 550, 2nd edn. Oxford: Oxford University Press, (1818) 1973.

Shove, Elizabeth. *Comfort, Cleanliness and Convenience: The Social Organization of Normality.* Oxford: Berg, 2003.

Shove, Elizabeth, Matthew Watson, Martin Hand and Jack Ingram. *The Design of Everyday Life.* Oxford: Berg, 2007.

Showers, Kate B. 'Beyond Mega on a Mega Continent: Grand Inga on Central Africa's Congo River'. In *Engineering Earth: The Impacts of Megaengineering Projects*, edited by Stanley D. Brunn, 1651–79. Dordrecht: Springer, 2011.

Shih, Shu-mei. 'Comparative Racialization: An Introduction'. *PMLA* 123, no. 5 (2008): 1347–62.

Silko, Leslie Marmon. *Almanac of the Dead.* New York: Penguin, 1992.

Simone, AbdouMaliq. *For the City Yet to Come: Changing African Life in Four Cities.* Durham, NC: Duke University Press, 2004.

Sipes, Todd. *Urban Exploration Photography: A Guide to Creating and Editing Images of Abandoned Places.* San Francisco: Peachpit Press, 2015.

Smith, Rick, and Bruce Lourie. *Slow Death by Rubber Duck: The Secret Danger of Everyday Things.* Berkeley: Counterpoint, 2009.

Soentgen, Jens. 'Die Bedeutung des indigenen Wissens für die Geschichte des Kautschuks'. *Technikgeschichte* 80 (2013) H4: 295–324.

Soentgen, Jens. 'Materialität'. In *Handbuch Materielle Kultur: Bedeutungen, Konzepte, Disziplinen*, edited by Stefanie Samida, Manfred K. Eggert and Hans Peter Hahn, 226–9. Stuttgart: J.B. Metzler, 2014.

Solnit, Rebecca. *Savage Dreams: A Journey into the Hidden Wars of the American West.* Berkeley: University of California Press, 1999.

Sperling, Julie. 'How They'll Know We Were Here: Plastic, Concrete, Aluminum'. Sperlingmosaics.com, 2018. Available online: https://sperlingmosaics.com/2018/01/how-theyll-know-we-were-here-plastic-concrete-aluminum/ (accessed 22 March 2023).

Spivak, Gayatri Chakravorty. *A Critique of Postcolonial Reason: Toward a History of the Vanishing Present.* Cambridge: Harvard University Press, 2003.

Stacks, Geoffrey. 'A Defiant Cartography: Linda Hogan's *Solar Storms*'. *Mosaic* 43, no. 1 (2010): 161–76.
Steffen, Will, Åsa Persson, Lisa Deutsch, Jan Zalasiewicz, Mark Williams, Katherine Richardson, Carole Crumley, Paul Crutzen, Carl Folke, Line Gordon, Mario Molina, Veerabhadran Ramanathan, Johan Rockström, Marten Scheffer, Hans Joachim Schellnhuber and Uno Svedin. 'The Anthropocene: From Global Change to Planetary Stewardship'. *AMBIO: A Journal of the Human Environment* 40, no. 7 (2011): 739–61.
Stoler, Ann Laura. 'Imperial Debris: Reflections on Ruins and Ruination'. *Cultural Anthropology* 23, no. 2 (2008): 191–219.
Stoler, Ann Laura. 'Introduction. "The Rot Remains": From Ruins to Ruination'. In *Imperial Debris: On Ruins and Ruination*, edited by Ann Laura Stoler, 1–38. Durham, NC: Duke University Press, 2013.
Stratford, Elaine, Godfrey Baldacchino, Elizabeth McMahon, Carol Farbotko and Andrew Harwood. 'Envisioning the Archipelago'. *Island Studies Journal* 6, no. 2 (2011): 113–30.
Sze, Julie. 'Boundaries of Violence: Water, Gender and Globalization at the US Borders'. *International Feminist Journal of Politics* 9, no. 4 (2007): 475–84.
Tait, Theo. '*How to Get Filthy Rich in Rising Asia* by Mohsin Hamid – Review', Guardian.com, 28 March 2013. Available online: https://www.theguardian.com/books/2013/mar/28/how-get-filthy-hamid-review (accessed 28 June 2018).
Taussig, Michael. *My Cocaine Museum*. Chicago: University of Chicago Press, 2004.
Taylor, Charles. 'Two Theories of Modernity'. In *Alternative Modernities*, edited by Dilip Parameshwar Gaonkar, 172–96. Durham, NC: Duke University Press, 2001.
Te Punga Somerville, Alice. 'The Great Pacific Garbage Patch as Metaphor: The (American) Pacific You Can't See'. In *Archipelagic American Studies*, edited by Brian Russell Roberts and Michelle Ann Stephens, 320–38. Durham, NC: Duke University Press, 2017.
Thompson, Michael. 'Rubbish Theory: The Creation and Destruction of Value', *Encounter*, June 1979. Available online: http://www.unz.com/print/Encounter-1979 jun-00012/?View=PDF (accessed 27 June 2019).
Thompson, Richard C. 'Plastics, Environment and Health'. In *Accumulation: The Material Politics of Plastic*, edited by Gay Hawkins, Jennifer Gabrys and Mike Michael, 150–68. London: Routledge, 2013.
Tillett, Rebecca. 'Introduction: Almanac Contextualized'. In *Howling For Justice: New Perspectives on Leslie Marmon Silko's Almanac of the Dead*, edited by Rebecca Tillett, 5–13. Tucson: University of Arizona Press, 2014.
Tweedie, Neil. 'Is the World Running Out of Sand? The Truth Behind Stolen Beaches and Dredged Islands', *The Guardian*, 1 July 2018. Available online: https://www.theguardian.com/global/2018/jul/01/riddle-of-the-sands-the-truth-behind-stolen-beaches-and-dredged-islands (accessed 5 June 2019).
Urry, John. *The Tourist Gaze*, 2nd edn. London: Sage, 2002.

van Mensvoort, Koert. 'Plastic Planet', Nextnature.com, 19 February 2011. Available online: http://www.nextnature.net/2011/02/plastic-planet/ (accessed 17 June 2018).
van Oss, Hendrik G. 'Background Facts and Issues Concerning Cement and Cement Data', *U.S. Geological Survey*, Open-File Report 2005-1152. Available online: https://pubs.usgs.gov/of/2005/1152/ (accessed 26 June 2016).
van Sebille, Eric, Chris Wilcox, Laurent Lebreton, Nikolai Maximenko, Britta Denise Hardesty, Jan A. van Franeker, Marcus Eriksen, David Siegel, Francois Galgani and Kara Lavender Law. 'A Global Inventory of Small Floating Plastic Debris'. *Environmental Research Letters* 10, no. 12 (2015): 124006.
VDZ. 'Global Cement Production from 1990 to 2030 (in Million Metric Tons)', Statista – The Statistics Portal. Available online: www.statista.com/statistics/373845/global-cement-production-forecast/ (accessed 28 November 2018).
Viney, William. '"A Fierce and Wayward Beauty": Waste in the Fiction of J.G. Ballard, Partis I & II', Ballardian, 11 December 2007. Once available online http://www.ballardian.com/a-fierce-and-wayward-beauty-parts-1-2 (accessed 27 June 2019).
Vitousek, Peter, and Oliver Chadwick. 'Pacific Islands in the Anthropocene'. *Elementa: Science in the Anthropocene*, 4 December 2013. doi: 10.12952/journal.elementa.000011.
Volz, Wolfgang, and Peter Pachnicke, eds. *Christo: Big Air Package*. Cologne: Taschen, 2013.
Vonnegut Jr., Kurt. *Slaughterhouse-Five, or The Children's Crusade*. New York: Bantam, (1969) 1994.
Waterman, Jonathan. 'Where the Colorado Runs Dry', *New York Times*, 14 February 2012. Available online: https://www.nytimes.com/2012/02/15/opinion/where-the-colorado-river-runs-dry.html (accessed 6 June 2018).
Waterman, Jonathan. *Running Dry: A Journey from Source to Sea Down the Colorado River*. Washington: National Geographic Books, 2010.
Waters, Colin N., Jan Zalasiewicz, Colin Summerhayes, Anthony D. Barnosky, Clément Poirier, Agnieszka Gałuszka, Alejandro Cearreta, Matt Edgeworth, Erle C. Ellis, Michael Ellis, Catherine Jeandel, Reinhold Leinfelder, J. R. McNeill, Daniel deB. Richter, Will Steffen, James Syvitski, Davor Vidas, Michael Wagreich, Mark Williams, An Zhisheng, Jacques Grinevald, Eric Odada, Naomi Oreskes and Alexander P. Wolfe. 'The Anthropocene is Functionally and Stratigraphically Distinct from the Holocene'. *Science* 351, no. 6269 (2016): 137–48.
Wenk, Silke. 'Bunkerarchäologien: Zur Einführung'. In *Erinnerungsorte aus Beton: Bunker in Städten und Landschaften*, edited by Silke Wenk, 15–37. Berlin: Links, 2001.
Whitlock, Gillian. *Postcolonial Life Narratives: Testimonial Transactions*. Oxford: Oxford University Press, 2015.
Whyte, Kyle Powys. 'Our Ancestor's Dystopia Now: Indigenous Conservation and the Anthropocene'. In *The Routledge Companion to the Environmental Humanities*,

edited by Ursula Heise, Jon Christensen and Michelle Niemann, 206–14. London: Routledge, 2017.

Willacy, Mark. 'A Poison In Our Island', Abc.com.au, 26 November 2017. Available online: http://www.abc.net.au/news/2017-11-27/the-dome-runit-island-nuclear-test-leaking-due-to-climate-change/9161442 (accessed 22 June 2018).

Woods, Derek. 'Scale Critique for the Anthropocene'. *Minnesota Review* 83 (2014): 133–42.

Wright, Alexis. *Carpentaria*. Artarmon: Giramondo, 2007.

Wu Ming-Yi. *The Man with the Compound Eyes*, trans. Derryl Sterk. London: Vintage, 2014.

Yablon, Nick. *Untimely Ruins: An Archaeology of American Urban Modernity, 1819–1919*. Chicago: University of Chicago Press, 2009.

Yaeger, Patricia. 'Introduction: Dreaming of Infrastructure'. *PMLA* 122, no. 1 (2007): 9–26.

Yamashita, Karen Tei. *Through the Arc of the Rain Forest*. Minneapolis: Coffee House Press, 1990.

Yarsley, Victor E., and Edward G. Couzens. *Plastics*. Harmondsworth: Penguin, 1941.

Young, Robert J. C. *Postcolonialism: A Very Short Introduction*. Oxford: Oxford University Press, 2003.

Zalasiewicz, Jan. 'A History in Layers'. *Scientific American* 315, no. 3 (2016): 30–7.

Zalasiewicz, Jan, Will Steffen, Reinhold Leinfelder, Mark Williams and Colin Waters. 'Petrifying Earth Process: The Stratigraphic Imprint of Key Earth System Paramters in the Anthropocene'. *Theory, Culture & Society* 34, nos. 2–3 (2017): 83–104.

Zalasiewicz, Jan, Colin N. Waters, Anthony D. Barnosky, Alejandro Cearreta, Matt Edgeworth, Erle C. Ellis, Agnieszka Gałuszka, Philip L. Gibbard, Jacques Grinevald, Irka Hajdas, Juliana Ivar do Sul, Catherine Jeandel, Reinhold Leinfelder, J. R. McNeill, Clément Poirier, Andrew Revkin, Daniel deB Richter, Will Steffen, Colin Summerhayes, James P. M. Syvitski, Davor Vidas, Michael Wagreich, Mark Williams and Alexander P. Wolfe. 'Colonization of the America, "Little Ice Age" Climate, and Bomb-Produced Carbon: Their Role in Defining the Anthropocene'. *The Anthropocene Review* 2, no. 2 (2015): 117–27. doi: 10.1177/2053019615587056

Websites without authors

'50th Anniversary of Nubia Campaign', UNESCO (United Nations Educational, Scientific and Cultural Organisation). Available online: http://whc.unesco.org/en/news/497/ (accessed 7 June 2018).

'Anthroslam'. *The Anthropocene Slam: A Cabinet of Curiosities*. Once available online https://che.nelson.wisc.edu/anthroslam/ (accessed 14 September 2018).

'Christo and Jeanne-Claude FAQ', Christojeanneclaude.net. Available online: http://christojeanneclaude.net/faq (accessed 28 June 2018).

Policy documents

EBRD. 'Chernobyl's New Safe Confinement', European Bank for Reconstruction and Development ('EBRD') (n.d.). Available online: https://www.ebrd.com/what-we-do/sectors/nuclear-safety/chernobyl-new-safe-confinement.html (accessed 27 June 2019).

IEA and the World Bank. 'Sustainable Energy for All 2015 – Progress Toward Sustainable Energy', International Energy Agency (IEA) and the World Bank. Washington: World Bank. doi: 10.1596/978-1-4648-0690-2. Available online http://hdl.handle.net/10986/22148 (accessed 12 October 2023).

WCD. 'Dams and Development: A New Framework for Decision-Making', World Commission on Dams. London: Earthscan, 2000. Available online: https://archive.internationalrivers.org/sites/default/files/attached-files/world_commission_on_dams_final_report.pdf (accessed 27 March 2023).

World Bank. 'Progress toward Sustainable Energy 2015', International Bank for Reconstruction and Development/The World Bank and the International Energy Agency. Washington: World Bank, 2015. Available online: doi: 10.1596/978-1-4648-0690-2.

Full-length/Feature films

Addicted to Plastic. [Film] Dir. Ian Connacher, Canada: Bullfrog Films, 2008.
High-Rise. [Film] Dir. Ben Wheatley, UK: Recorded Picture Company, 2015.
Operation Glen Canyon. [Film] United States Department of the Interior Bureau of Reclamation (USBR), 1961.
Plastic Planet. [Film] Dir. Werner Boote, Germany/Austria: Brandstorm Entertainment, 2009.
Supersize Me. [Film] Dir. Morgan Spurlock. USA: Daryl Isaacs, 2004.

Short films

1953 Advertisement for Saran-Wrap, Archive.org. Available online: https://archive.org/details/1953CommercialForSaran-wrapad2 (accessed 28 June 2018).

'The Cracking of Glen Canyon Damn with Edward Abbey and Earth First!' Available online: http://www.sacredland.org/the-cracking-of-glen-canyon-damn-with-edward-abbey-and-earth-first/trackback/ (accessed 6 June 2018). Produced by Christopher (Toby) McLeod, Glenn Switkes and Randy Hayes.

Television series

Sesame Street. Created by Joan Ganz Cooney and Lloyd Morrisett. Dir. diverse. 1969–present.
The Knick. Created by Jack Amiel and Michael Begler. Dir. Steven Soderbergh. Cinemax, 2014–15.

Index

Abbey, Edward 92–3, 96–7
  The Monkey Wrench Gang 84, 93, 97–8, 99–100, 113
Abu Simbel 92–3, 108–9
Adams, Ansel 87–9
  'Boulder Dam 1941' 89
  'Boulder Dam 1942' 87
affect 64, 79 n.25, 86, 120–1, 150 n.4, 159, 175–6
affluence, material 141
agency 18–20, 32, 35, 45, 68, 104, 118–19, 158. *See also* agential realism, intra-action, material agency
  creative 21
  cultural 9
  distributed 86
  of place 164
agential realism (Barad) 83
Ahmed, Sara 150–1, 150 n.4
Alaimo, Stacy
  trans-corporeality 4–5, 5 n.4, 49–50, 55–6
Alliance of Small Island States 57
alterlives (Murphy) 184–5
Anthropocene 7, 27, 28, 32–5, 184, 186
  and colonization 33
  discourses of 32–3, 39
  markers of 33–5, 34 n.36, 36
  and radionuclides 33
  scales of the 186–7
  spatiality 36–7
  symbol of 46
  temporality 36
anti-tourism 161, 169
archaeology 41, 164, 171. *See also* future archaeologist
archipelagic thought 56, 58 n.9, 79–80, 79 n.25, 81. *See also* epistemologies, Santos Perez
archipelago 56, 79–80, 79 n.25, 81
archive 9, 63–4, 164, 168, 168 n.17, 185

artefact 4, 8, 21, 33, 37, 38, 39, 40, 120–1, 190. *See also* future artefact

Ballard, J. G. 164
  Concrete Island 38
  High-Rise 172–8
Barad, Karen 9 n.9, 36, 37, 46
  agential realism 83–4
  diffraction 9, 47–8, 93–4
  intra-action 20, 43, 47, 105, 106, 159, 166
Barthes, Roland 13, 78–9, 119
beachcombing 61–2, 66
Bennett, Jane 44–5, 86
  thing-power 42–3
  vibrant matter 42, 62–3, 86, 164
Berlin Wall 1–2, 3, 7–8, 128 n.9, 188
Bhabha, Homi 98, 98 n.19
Bhopal disaster 122–3
biodegradation 13 n.17, 59–60, 118, 145–6, 181–2
bioplastic. *See* plastic
birds
  duck (*see* rubber duck)
  extinct species 55
  Laysan albatross 59, 60–1, 61 n.11, 62, 65–6 (*see also* Jordan, Liittschwager, Middleton)
bisphenol A (BPA) 6, 30–1, 69, 71–2
Blackwell, Andrew
  Visit Sunny Chernobyl: Adventures in the World's Most Polluted Places 58–9, 169–70
body 42, 68, 72, 156, 163 n.14, 176
  corporate 68
  as material form 43–4
  medical 137–8, 140
  and ruins 148–9, 157–8
  sensory 4–5
  social 86
  and toxicity 77
  trans-corporeality 5 n.4

Boote, Werner
  Plastic Planet 115–16, 116
bottles 31, 56, 71, 117, 124 n.6, 141, 142–3, 181–2
Brown, Bill
  thing theory 3
bunker 147, 164, 175. *See also* Ballard

Carle, Eric
  10 Little Rubber Ducks 55, 66, 73
Carson, Rachel
  Silent Spring 55
Cartesian dualism 41
cellophane 74
cement 8, 16, 24, 25, 34, 84–5, 103, 177, 182. *See also* concrete
  discovery of 17–18
  global dimension 30
  history of 13–14
  technology of 14
Chakrabarty, Dipesh 32
Chernobyl 77–8, 168–70. *See also under* nuclear
  museum 169
  Sarcophagus 169–70
Christo and Jeanne-Claude 117, 118, 125–31
  borders 128–9
  fabric 128–9, 131
  'Floating Piers' 129, 130–1
  liminality 128, 129
  'Ocean Front' 129
  'The Pont Neuf Wrapped (1975–85)' 129
  'Running Fence Project' 128
  'Surrounded Islands, Biscayne Bay, Greater Miami, Florida, 1980–83' 126–8
  'Wrapped Coast' 129
climate change 34, 80, 91 n.6, 170, 184
cling-wrap 117, 118, 119–25, 145, 183, 184
  advertising of 119–20
  affect 120
  displacement 120
  histories of 121, 122–3
  and objectification 120–1
  temporality 117, 123–5
  toxicity of 122 (*see also* Bhopal disaster)
  visuality of 120
colonialism 36–7, 49, 85, 92, 106, 107, 108, 110–11
colonial modernity 26–7
concatenation of plastic 65, 71, 74–5, 79, 183–4
concrete 1, 4, 5, 6, 11, 30, 84–5, 85–6, 104, 112, 152, 168–9, 173
  See also cement, materiality, material poetics, megadams, nuclear waste, plasticoncrete, ruins
  agency 159
  containment 168–9, 171–2, 177
  and cultural practice 17
  definitions of 11 n.13, 14
  displacement 177
  dust 156
  environment 177
  etymology 10–11
  fluidity 15–16, 114, 182
  global dimensions of 30, 31
  history of 13–14
  infrastructure 177
  invention 17–18
  inventory of 23–4
  malignant agency 149–50, 153, 155–6, 169–70, 175–6
  modernity 24–5, 151, 165–6
  and place 106–7
  Portland cement 17–18
  and power 167–8
  as process 14, 15, 16, 149
  referentiality 11
  replacement 181
  spatiality 14–15, 106
  technological practice 24–5
  temporality 14–15, 149, 165–6, 183–4
consumption 5, 27, 42, 56–7, 60, 64, 90, 142–3, 166
Coole, Diana 41–2, 43, 44
Couzens, Edward G.
  'Plastic Man' 12, 67, 79, 183–4

dams. *See* megadams
Davis, Heather 23 n.23, 38, 56–7, 59–60, 70, 115, 120–1, 144–5, 184–5
De Botton, Alain
  The Pleasures and Sorrows of Work 133–4, 135

debris 53 n.2, 57 n.6, 127–8, 147–8, 148, 151. *See also* ruin, waste
  marine 31 n.31, 56, 64–5, 74–5
  plastic 34, 61–2
DeLoughrey, Elizabeth 33, 57
  tidal dialectics 81
Denkobjekt 3, 60. *See also* Denkstoff, epistemologies
Denkstoff 3, 5, 8
development 20–1, 85–6, 92–3, 96, 105–6, 114, 141, 153, 162
Development-Induced Displacement and Resettlement (DIDR) 107, 109
diffractive methodology 9, 47–8, 93–4
discovery, narratives of 3, 17–20, 26–7, 122 n.5, 162, 164. *See also* material poetics
displacement 100–1, 105–6, 107, 120, 177
  of energy 84–5, 85–6
  of indigenous populations 94, 109, 110, 111
  and place 107
  and resettlement 107
documentary
  films 84, 96, 103
  photography 63–4
dystopia 36–7, 175–6, 186

Earth First! 84, 113
  'The Cracking of Glen Canyon Damn with Edward Abbey and Earth First!' 93, 96–7, 99–100, 101
earthquake 149–50, 151, 152–3, 154, 156
Ebbesmeyer, Curtis 66
  Flotsametrics and the Floating World 55
ecoterrorism 49, 97–8, 99 n.20, 101 n.22
eco-warriors 98–9, 100
Edensor, Tim 147–8, 156–9, 160–1
edgework 163, 166 n.16. *See also* Farley, Symmons Roberts
electricity 84–5, 90, 100. *See also* energy, power
emissions 34 n.36, 91, 177, 183–4
endocrine disrupting chemicals (EDCs) 15–16, 62, 76–7, 137
energy 89, 90, 91–2, 94, 104, 158–9
  displacement of 85
  emissions 91

entanglements 20–1, 27–8, 47–8, 104, 181, 184, 186
environmental ethics 59, 62–3, 70. *See also* ethics of story-telling
  and children 70
  and future artefact 70 (*see also* future artefact)
environmental justice 84, 188
epistemology 49–50, 115, 136
  indigenous 3, 19
  visual 4, 49–50, 107, 180 n.2
  Western 50
ethics of story-telling 151
ethnic politics 98

factories 154–7, 158
Farley, Paul
  Edgelands 160–1
Farmer, Jared
  future fossil 39–40
  Glen Canyon Dammed 95–6
filth 141–2
flotsam and jetsam 55, 64–5, 66, 67, 182–3
  distinction between 67–8
  narrativization of 67–8
fluidity 6, 29, 182
fossil 13, 16, 31, 38, 39, 51, 129–30, 185. *See also* future fossil, technofossil
Franklin River 84, 108
Freinkel, Susan 58
  Plastic: A Toxic Love Story 60–3, 116, 121, 122 n.5
Frost, Samantha 41–2, 43, 44
future
  archaeologist 37, 38, 164, 177–8, 186
  artefact 7, 35–40, 117, 162, 167, 170–1, 185, 190
  fossil 38–40 (*see also* fossil)
  geologist 38
  histories 117
  remains 38–40
  ruins 40, 156 (*see also* ruins)

Garrett, Bradley
  Explore Everything: Place-Hacking the City 162–4
genetic alteration 77
geology 32, 38, 185. *See also* Parikka

Glen Canyon Dam 83–4, 93–104. *See also* megadams
  Almanac of the Dead (Silko) 84, 93, 98–101
  'anti-dam activism' 95–6
  construction 102–4
  'The Cracking of Glen Canyon Damn with Edward Abbey and Earth First!' 96–7
  Glen Canyon Dammed (Farmer) 95–6
  The Monkey Wrench Gang (Abbey) 84, 93, 97–8, 99–100, 113
  'Operation Glen Canyon' 101–2
  and recreation 95–6
  site 94
globalization 15, 30 n.29, 36, 182 n.5
Grand Inga project 90. *See also* megadams
Great Pacific Garbage Patch (GPGP) 53–4, 58, 66, 79–80, 81, 138 n.14
Greenland 76–7
Green Party of Australia 84
Gunesekera, Gomesh
  Reef 8

Hamid, Mohsin
  How to Get Filthy Rich in Rising Asia 117–18, 135, 139–45
Hawkins, Gay 21, 31, 123–4, 142–3
heritage site 157 n.7, 160–1, 164
High Aswan Dam 108. *See also* megadams
Hogan, Linda
  Solar Storms 47, 48–9, 50–1, 99 n.20
Hohn, Donovan
  Moby-Duck: The True Story of 28,800 Bath Toys Lost at Sea 66–8, 73
hydropower 50–1, 84, 89, 91, 94, 110. *See also* electricity, energy, megadams
hygiene 12, 67, 136, 141–2. *See also* plastic practices
hyperaesthetics 166, 167

illness 117–18, 135, 139–40. *See also* body
imagery 49–50, 58–9. *See also* material poetics, visual representation
indigenous
  displacement 110–11
  epistemologies 3, 98 (*see also* epistemologies)
  knowledge 19

Native Americans 49
  populations 77
  use of materials 19
industrial production 29, 181
industrial ruins. *See* ruins
infrastructure 6, 10, 10–20, 23–4, 83–4, 87–9, 98, 104, 177, 180
  medical 140
  nuclear 167–8
  poetics of 84 (*see also* material poetics)
  and power 87–8
interconnectivity 6, 67, 135, 138, 140
  and space 134–5
intra-action (Barad) 20, 43, 47, 105, 159. *See also* agency, agential realism
invention 8, 17–24, 37, 40, 46, 53, 185, 189–90
  as concept 20
  narratives of 20–1, 24 (*see also* material poetics)
inventory 23–4, 71, 131–2
  etymology of 22
irrigation 89, 108
island 53–4, 56, 57, 58, 79–80, 81, 126–8, 129, 170. *See also* archipelago, Christo and Jeanne Claude

Jetñil-Kijiner, Kathy
  'Anointed' 170
Jordan, Chris
  'Midway: Message from the Gyre' 63–4

King, Thomas
  The Inconvenient Indian: A Curious Account of Native People in North America 84, 105, 110–11
Kippfigur 43
Klein, Naomi
  'The Right to Regenerate' 70

Laferrière, Dany
  The World is Moving Around Me: A Memoir of the Haiti Earthquake 149–54, 177
landscape 43, 87–8, 96–7, 99, 101–2, 151–2, 161, 175–6
Latour, Bruno 20 n.22
Laysan albatross. *See* birds, Jordan, Liittschwager, Middleton

Le Corbusier 16–17, 25, 175
Liboiron, Max 6, 20 n.22, 27 n.25, 30 n.30, 55 n.5, 70 n.19, 72, 106, 124–5
Liittschwager, David
　'Shed Bird' 64–5
Lin 'Cypherone' Yung-Jie
　'Abandoned Future IV' 164–5
livelihoods 78, 101–2, 105–6, 109, 114, 134, 159
logic of externalization 184
logistics 15, 24, 30, 117, 133, 140, 142, 181
Lourie, Bruce
　Slow Death by Rubber Duck: The Secret Danger of Everyday Things 55–6, 69–70, 71–2, 137–8

magic (Taussig) 13
Māori ontology 3. *See also* indigenous epistemologies
marine debris. *See* debris
Marriott, James 117, 132–4
material
　methods 46–7, 189
　processes 14–15, 34, 85
　properties 5
　relations 8, 21, 43, 46–52, 56, 158, 176
　thinking 60, 187
material agency 34, 41–3, 45, 122 n.5, 158. *See also* agency
　agential realism 83
　concrete 148–9, 149–50, 153, 155–6, 175–6
　distributed 86
　and invention 20
　malignant 62, 149–50
　plastic 75, 79, 81, 130–1, 145–6
　of things 41
　toxic 72, 77, 81
material culture 5, 44, 101, 162
Material Culture Studies 40–2
material ecocriticism 41–2, 43–4, 101
materiality 3–5, 8–9, 28–9, 44–5, 88–9. *See also* concrete, material, material poetics, plastic
　buildings 173
　petrochemicals 44
　relationality of 5, 131–2
　scales of 6–7
　and social relations 176

and space 4
and time 4, 8
and visuality 4, 110 n.28
material poetics 50, 143–5, 172–4
　concrete 150–1, 175–6
　discovery narratives 17–20
　imagery 49–50
　of infrastructure 84
　invention narratives 20–1, 24
　mimesis 9, 38, 47–8, 49–50, 104, 112–13
　narrativization of flotsam and jetsam 67–8
　plastic narrativity 132–3, 135
　repetition 112–13
　replacement narratives 181–2
　risk and narrativity 73
　vignettes 150–3
Matilija Dam 113. *See also* megadams
megadams 83–114. *See also* Glen Canyon Dam
　'Boulder Dam 1941' (Adams) 89
　'Boulder Dam 1942' (Adams) 87
　'The Cracking of Glen Canyon Damn with Edward Abbey and Earth First!' 96–7
　and development 85, 92–3
　displacement 85, 107, 109, 110, 111 (*see also* displacement, indigenous populations)
　ecoterrorism 97–8
　and environmental justice 84
　externalization of land 106
　Franklin River 84
　Grand Inga project 90
　green energy 91
　High Aswan Dam 108
　The Inconvenient Indian: A Curious Account of Native People in North America (King) 110
　indexicality of 85, 114
　infrastructure 83–4, 87–8
　The Monkey Wrench Gang (Abbey) 84, 93, 97–8, 99–100, 113
　and nation-building 110
　Navajo Generating Station 94 (*see also* electricity, megadams power)
　Nubia Campaign 108
　An Ordinary Person's Guide to Empire (Roy) 111

and place 106–7
and power 83, 84, 87, 88–9, 105, 111–12 (*see also* electricity, energy, terraforming)
as sacred place 92–3
and slow violence (Nixon) 105
World Commission of Dams 84, 92
Marriott, James 117
  'Where Does This Stuff Come From? Oil, Plastic and the Distribution of Violence' 132–4, 135
Massey, Doreen
  space 107, 134–5
  stories-so-far 93, 117–18, 134–5, 137, 151
microplastics. *See* plastic
Middleton, Susan
  'Shed Bird' 64–5
migration 106
mimesis. *See* material poetics
Minio-Paluello, Mika 117
  'Where Does This Stuff Come From? Oil, Plastic and the Distribution of Violence' 132–4, 135
Modernism 24–5, 26
modernity 24–8, 98, 110, 114, 135–6, 142–3, 144–5, 189. *See also* Modernism
  accultural 26
  and colonialism 26
  institutions of 151–2
  linear trajectory 19–20
  material 27
  medical 137–8, 139, 140–1
  patternings of 27
  plastic 25
  plurality of 136–7
  processes of 145
  progressive 25
  and waste 56–7
molecular scale 6, 15–16
Moore, Charles
  Plastic Oceans 63, 74–7, 182–3
Motlagh, Jason
  'The Ghosts of Rana Plaza' 155–6
museum 40, 53 n.1, 108–9, 125–6. *See also under* Chernobyl

narrative. *See* material poetics
nation 49, 94, 106, 109, 152
building 110 (*see also* unimagined community)
Native Americans 49, 94, 95, 110–11. *See also* indigenous populations
natureculture 88–9
Navajo Generating Station 94, 95. *See also* electricity, energy, megadams, power
neo-imperial project 106
nomadism 108–9, 114
Nubia Campaign 108–9. *See also* megadams
Nubian people. *See* nomadism
nuclear. *See also* Chernobyl
  infrastructures 167–8
  Pacific (DeLoughrey) 33, 57
  power 167–8
  testing 57
  toxicity 77–8
  waste 168–9, 170–2
Nylon 20–1, 130–1

object-ness 3
  and sensuality 4–5
  and visuality 4
obsolescence (planned) 15. *See also* waste
Onkalo 171–2. *See also* nuclear waste

Pacific Ocean 53, 74–5, 78. *See also* Plastic Pacific
Parikka, Jussi 9, 14
  A Geology of Media 37–8, 40, 185–6
Phillips, Cassandra
  Plastic Oceans 63, 74–7, 182–3
photographic editing 166–7. *See also* photography, hyperaesthetics
photography 4, 63–4. *See also* photographic editing, hyperaesthetics
phthalates 69–70
place 106–7
  relationality of 106–7
  and space 107
plastic 4, 5–7, 10, 11–14, 15–17, 56, 58, 115, 119, 135–7, 140–1, 142, 179–84
  agency 62, 75, 81 (*see also* material agency)
  bioplastic 13 n.17
  cling-wrap (*see* cling-wrap)
  concatenation of 61, 63, 65

as cultural practice 11–12, 16
debris 61–2
decomposition 59–60
discovery of 18–19, 122 n.5
etymology 11, 11 n.15
fabric 128–30
forms of 22–3
global dimensions of 30–1
high performance 22–3
history of 13, 122
inventory of 22–3
materiality of 74–5
microplastic 76 n.22, 179 n.1
and modernity 25
narrative 132–3, 135
nurdles 76
packaging 145
pollution 77, 136
as process 15
relationality 59–60, 73
technical co-evolution 21
temporality 59–60, 124–5, 126, 184–5
toxicity 55, 60, 75, 76–7, 81
trans-corporeality 65–6
waste 53, 55–7, 58, 59, 66, 115–16, 116 (*see also* waste)
plasticity 4, 31, 46, 47, 115, 132–3, 135, 182, 183. *See also* Roberts
epistemic 115
plastic practices
medical 137–9, 139–41
wrapping as 117, 118, 126–7 (*see also* Christo and Jeanne Claude, cling-wrap)
plasticoncrete 6–7, 10–15, 24, 33, 34, 177–8, 188
fluidity 15
heuristics of 10, 181, 181–2, 183
as process 15 (*see also* material processes)
Plastic Pacific 53–7, 58–9, 60, 67–8, 74, 79–80, 81, 179–80, 182–3, 186–7
See also Pacific Ocean
Plastic Age 12, 79, 183–4
plastiglomerate 33–4, 158, 178. *See also* plasticoncrete
discovery/invention 53–4
Plumwood, Val 9, 187 n.10
poetics. *See* material poetics

pollution 75, 76, 77, 78, 136, 188. *See also* toxicity, waste
of bodies 76–7, 138
indices of 77
polychlorinated biphenyls (PCB) 6, 62, 76–7
Potter, Emily 21, 31, 123–4, 142–3
power 83–4, 87–8, 89–90, 111–12, 147–8, 167–8
colonial 49, 108
discursive 32–3, 100–1
and infrastructure 87–9 (*see also* electricity, infrastructure)
nonhuman 42 (*see also* material agency, thing power)
resistance to 49
Pratt, Mary Louise
re-invention 21–2
preservation 120–1, 124–5, 126, 131, 145, 164
provenance stories 117, 131–4, 189, 190. *See also* future artefact, future histories
trajectories of 134

Race, Kane 21, 31, 123–4, 142–3
Rana Plaza collapse 154–5, 156
recycling 31, 80, 127–8, 182
reflective methodology 47–8
refugees 131
re-invention (Pratt) 21–2
replacement narratives 181–2. *See also* material poetics
retrofuture 165
risk 12, 67–8, 69
externalization 22
health 116–17, 157–8
isolation of 73
and narrativity 72–3
and toxicity 59, 72–3
Roberts, Jody A.
'Reflections of an Unrepentant Plastiphobe: An Essay on Plasticity and the STS Life' 25, 117–18, 135, 137–9
Romantic tradition 147, 161–2
Roy, Arundhati 84, 109, 110, 112–13
Listening to Grasshoppers: Field Notes on Democracy 111–12

'Mr Chidambaram's War' 114
An Ordinary Person's Guide to Empire 112
'Public Power in the Age of Empire' 111–12
rubber 19
rubber duck 53–4, 55–6, 58–9, 66–73
  bisphenol A (BPA) 71–2
  children and 67, 70–1
  cultural significance of 67
  ethics 70
  Flotsametrics and the Floating World (Ebbesmeyer, Scigliano) 55
  iconicity of 69, 70
  invention of 67
  Moby-Duck: The True Story of 28,800 Bath Toys Lost at Sea (Hohn) 66–8, 73
  Slow Death by Rubber Duck: The Secret Danger of Everyday Things (Smith, Lourie) 55–6, 69–70, 71–2, 137–8
  10 Little Rubber Ducks (Carle) 55, 66, 73
  toxicity of 69–70 (*see also* phthalates)
Rubber Duck Wars 69–70, 71
ruins 38–9, 40, 147–9, 167. *See also* future artefact, future fossil, future remains, nuclear
  and body 157–8
  and economic value 159
  heritage 161
  and imperialism 148–9
  industrial 156–7, 158
  materiality of 157, 158, 159
  nuclear 167–72
  and photographic practices 161 (*see also* photographic editing, photography)
  political dimension 147–8
  as process 157 n.7, 158–9, 162–3
  reappropriation 158–9
  as slow violence 156
  social 176
  space 157–8, 159
  temporality 156–7, 158, 159
  toxicity of 170
  and urban exploration 160–7, 177–8
Runit Island 170, 171, 171–2

St. John Mandel, Emily
  Station Eleven 40, 189

Santos Perez, Craig 56, 78
  'Age of Plastic' 56, 78–80 (*see also* Great Pacific Garbage Patch)
  archipelagic thought 79–80, 81
Sanzhi Pod City 164–5
Saran Wrap. *See* cling-wrap
Scigliano, Eric
  Flotsametrics and the Floating World 55
Sedgwick, Eve Kosofsky 28
sediment 11 n.13, 36, 38, 93, 95
Silko, Leslie Marmon
  Almanac of the Dead 84, 93, 98–101
Small Island Developing States (SIDS) 57
Smith, Rick
  Slow Death by Rubber Duck: The Secret Danger of Everyday Things 55–6, 69–70, 71–2, 137–8
souvenir 1–4, 5, 7–8, 188
  materiality of 4
Symmons Roberts, Michael
  Edgelands 160–1
stories-so-far (Massey) 93, 107, 117–18, 134–5, 151

technical co-evolution 21
technofossil 31, 185 (*see also* fossil, future artefact)
Te Punga Somerville, Alice 56, 81
thing. *See also* thing-power
  definition of 3, 3 n.3
  relationality 3
  social life 41–2
thing-power (Bennett) 42–3
thing theory (Bill Brown) 3
tidal dialectics (DeLoughrey) 81. *See also* epistemologies
toxicity 60, 62, 69–70, 76–8, 170
  children 71
  communication of 171–2
  containment 171
  icon 55
  indices of 77–8
  marine 76
  and pollution 77
  relations of 81
  and risk 72–3
trans-corporeality 4–5, 5 n.4, 49–50, 60, 65–6, 159
transmission 56–7, 184–5

unimagined communities (Nixon) 102–3, 105, 109, 110–11, 112
United States Department of the Interior Bureau of Reclamation 84, 93
 'Operation Glen Canyon' 101–4
urban exploration (urbex) 160–7
 and agency of place 164
 definition of 161–3, 161 n.11
 and discourses of discovery 162
 and future artefacts 162, 164
 and nostalgia 164 (*see also* edgework)
 photographic practices of 161, 162–3, 164–5, 166–7 (*see also* photography and editing)
 practitioners 163 n.14
utopia 165–6

vibrant matter. *See* Bennett, vital materialism
violence 6, 32, 49, 117, 132, 155, 173
 slow 38–9, 77–8, 105, 156, 188
viscose 18–9
visual representation 4, 49–50, 58–9, 60, 74–5, 102, 103. *See also* material poetics
vital materialism (Bennett) 42–3, 62–3, 76–7, 86

Vonnegut Jr., Kurt
 Slaughterhouse Five 189

waste 30–1, 53–4, 56–7, 66, 125, 134. *See also* debris
 creation of collective 15
 export of plastic 148
 medical 138
 nuclear waste 168–9, 170–2
 ruins and 147–8, 158
 social behaviours 174
 temporality 183–4
water 68, 83, 84–5, 90–1, 93–4, 94, 103–4, 105. *See also* fluidity, megadams, Pacific Ocean
 bottled 71–2, 141–3
 as motif 50–1
Western cultural dominance 26
World Bank
 'Progress toward Sustainable Energy 2015' 91–2
World Commission on Dams 84, 85, 92, 105–6

Yarsley, Victor E.
 'Plastic Man' 12, 67, 79, 183–4

www.ingramcontent.com/pod-product-compliance
Lightning Source LLC
Chambersburg PA
CBHW052108300426
44116CB00010B/1586